AF345644

Dr. rer. nat. habil. Steffen Sykora

Basics of Superconductivity

An Introduction to Theoretical Physics

© 2024 Steffen Sykora

Druck und Distribution im Auftrag des Autors:
tredition GmbH, Heinz-Beusen-Stieg 5, 22926 Ahrensburg,
Germany

ISBN
Paperback   978-3-384-35483-9
Hardcover   978-3-384-35484-6
e-Book   978-3-384-35485-3

Contents

Preface

Aim of the present textbook is to give a thorough introduction to important theories of superconductivity in such a way that the reader will be able to get a deep understanding of the prominent effects in superconductors. Particular importance is attached to explain the basic properties of superconductivity and its most important applications. Thereby, in the present textbook we focus on the phenomenological theories of superconductivity. This means that the theoretical concepts are based on a few assumptions about the superconducting state, which are motivated by experiments, without going into the details of the underlying microscopic system. This is completely sufficient for a good understanding of the phenomena as this book strives for. The treatment of the microscopic theory of superconductivity, in particular the theory of Bardeen, Cooper, and Schrieffer (BCS), is subject of an upcoming second part of this textbook which is planned to be published directly after the present textbook.

After explaining the fundamental assumptions in great detail, all physical properties and effects in superconducting systems are derived from that. Great importance is placed on a detailed and comprehensible derivation of all formulas. A brief introduction to important concepts of classical physics (mechanics, electrodynamics, thermodynamics) and to the basic idea of quantum mechanics is given. All definitions and basic assumptions are explained in great detail, and all theoretical results are derived from these basic equations in such a way as the reader is able to follow the particular steps without additional literature. Final aim is to give the reader the ability to understand the basic experimental properties of superconductors.

An overview of these experimental findings is given in chapter 1 in form of a rather compact illustration. The theoretical explanations of these findings are then presented in a comprehensive way in the subsequent chapters 2-5. The following topics will be dealt with: Chapter 2 develops the basic properties of the London theory of superconductivity. Particular attention is given to the explanation of the famous Meissner effect and the theoretical description of the magnetic field inside a superconductor.

Chapter 3 gives an introduction to the basics of thermodynamics, which is needed for the concepts in this textbook. Starting with the laws of thermodynamics, the general concept of thermodynamic potentials is introduced and free energy and Gibbs free energy are defined. In this context, the concept of functional derivation is also discussed. The chapter ends with a basic introduction to the Landau theory of phase transitions.

Chapter 4 concentrates on the theory of the phase transition from the superconducting state to the normal state within the Ginzburg-Landau theory. Starting from the Landau theory of phase transitions, the concepts of order parameters and the minimization of the Landau potential are described. These considerations lead to a basic set of field equations describing the superconducting state as well as the magnetic field inside the superconductor. The equations are extensively derived and solved for specific cases. In this context, the characteristic length scales determining the superconducting state are also discussed.

Finally, in chapter 5, prominent applications of the Ginzburg-Landau theory are presented. Particular attention is given to the critical magnetic field. In this context, the chapter develops a systematic access to type 1 and type 2 superconductors and provides the derivation of the characteristic parameters to distinguish the two types of superconductors. Moreover, the lower and upper critical values of the magnetic field in type 2 superconductors are explicitly calculated. The chapter provides a comprehensive description of Abrikosov vortices and their relevance for the field penetration.

Throughout this book, we shall use Gaussian units. Relevant equations within this textbook are numbered and referred in the text by citing the corresponding equation number. All figures are numbered as well.

Chapter 1 – Characteristic properties

The defining property of a superconductor is the disappearance of electrical resistance at very low temperatures. Associated with the infinitely good conductivity is the ability to completely displace a magnetic field from the interior of the material. The superconductivity is found in all metals and also in other compounds. Therefore, it is a fundamental property of condensed matter that goes far beyond a presence only in some special compounds. The phenomenon was discovered by Onnes in 1911.

In this chapter an overview of the known experimental properties of superconductors is given. Starting with the basic phenomena of the vanishing electrical resistance in section 1.1 and Meissner effect in section 1.2, all further important effects are described in the sections 1.3-1.6.

1.1 – Electrical resistance

In an ordinary metal at room temperature, the electrical resistance is relatively small, but not zero. Usually, one measures the resistance by applying a bias voltage U and measuring the associated current I. The electrical resistance R is then defined by the ratio $R = U/I$, since the Ohm's law $I \propto U$ applies to metals at room temperature. If the same measurement is carried out at an extremely low temperature below a certain critical temperature, the current becomes 'infinitely large', so that the value zero is assigned to the resistance. Then it is said from an experimental point of view that the material is superconducting.

The critical temperature T_c below which the resistance vanishes (usually called transition temperature) depends strongly on material properties and external parameters. One example of a quantity which controls the superconducting transition very well is an externally applied magnetic field.

Figure 1.1 shows the electrical resistance of a superconductor (solid line) at low temperatures as typically measured by experiments. The central observation is a sharp transition from a typical behavior of the resistance of normal metals to unmeasurable small values if the temperature is cooled down below T_c. For comparison, the temperature behavior of a usual metal, where the superconducting transition has been suppressed (for example using magnetic fields), is also shown (dashed line).

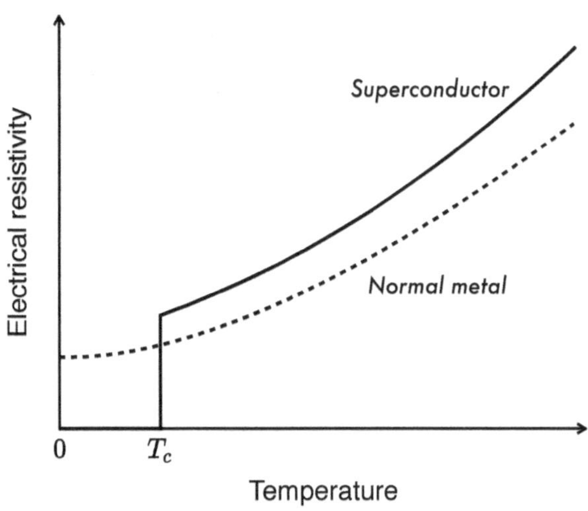

Figure 1.1: Schematic picture of the temperature behavior of a superconductor in comparison to a usual metal. Below a characteristic temperature the electrical resistance drops down to zero.

The property of the vanishing resistance was discovered during resistance measurements on mercury. The transition temperature for mercury is relatively small, $T_c = 4.2$ K, but the transition to a superconducting state has been found very soon also for other metallic compounds where the transition temperature might be slightly higher. Figure 1.2 shows measured values of transition temperatures for selected superconducting materials and the corresponding year of discovery. The conventional superconductors (circles) have T_c values up to ≈ 40 K for MgB_2. The well-known copper-based

10

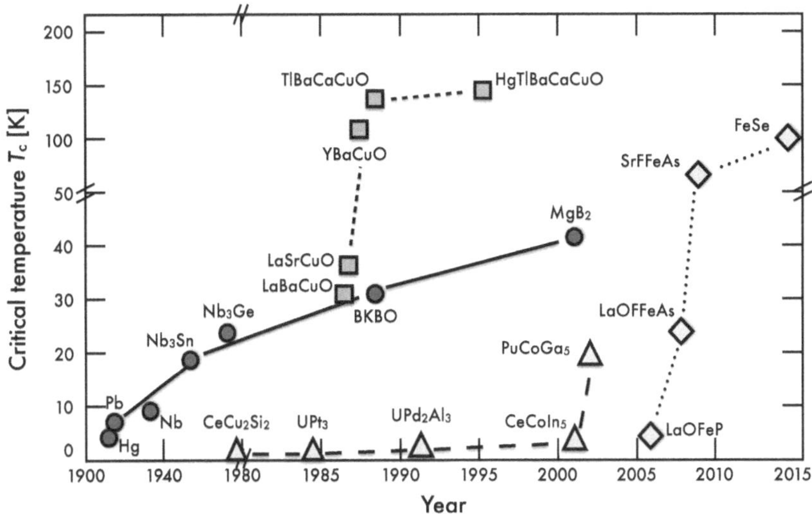

Figure 1.2: Timeline of the discovery of selected superconducting materials and the corresponding transition temperatures up to the year 2015. The conventional superconductors (circles) have transition temperatures up to 40 K. Higher values of T_c are possible for the copper-based superconductors (squares) and iron-based superconductors (diamonds). The heavy fermion superconductors (red triangles) are also counted to the unconventional superconductors despite their low T_c values.

high-temperature superconductors are characterized by relatively large critical temperatures up to 150 K at normal pressure.

Using a magnetic field which changes in time, it is possible to induce a steady current in a superconducting loop. This phenomenon is closely related to the vanishing electrical resistance. Experiments could not find any measurable reduction of the steady current over decades, i. e. the half-life is usually measured to be larger than 10^6 years.

1.2 – Meissner effect

If charges in the material can be displaced infinitely easily due to the lack of resistance, it is easy to imagine that they are also extremely sensitive to magnetic fields. The reason for this is that a small change in the magnetic field immediately leads to an induction of an electric field, to which the superconducting charges then react immediately with a large electric current. This current, in turn, generates a magnetic field that counteracts the external field. Experiments show that in most cases even the external field is completely displaced by this effect. This phenomenon, which occurs only in the superconducting state, is called the Meissner effect. It should be noted that metals in their normal state let the magnetic field almost completely into the material.

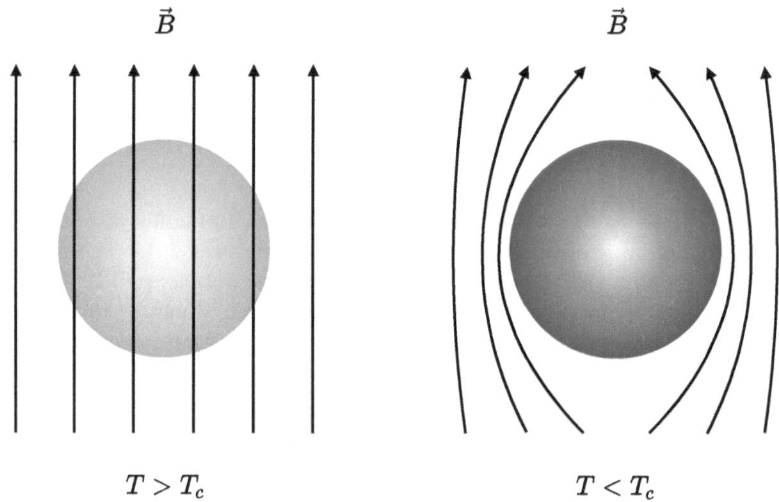

Figure 1.3: Meissner effect in a bulk superconductor. An external magnetic field $\vec{B}$ (field lines displayed by arrows) penetrates a metallic material (normal state, light gray). If the material is cooled down to temperatures below the transition temperature T_c (superconducting state, dark gray) the magnetic field is displaced so that the interior of the superconductor is field-free, $B = 0$.

Thus, as a consequence of the vanishing electrical resistance a superconductor strongly interacts with an external magnetic field. The effect is a displacement of the magnetic field from the interior of a superconductor during its transition to the superconducting state. A schematic picture of the Meissner effect is given in Figure 1.3. It was discovered in 1933 by Meissner and Ochsenfeld from measurements of the magnetic field distribution outside superconducting tin and lead samples.

The reaction of a material to an external magnetic field by the formation of its own magnetic field in the interior of the material is called *magnetization* of the material. The superconductor should thus have a very large magnetization directed against the external field. One speaks of ideal diamagnetism. As we will show in the following, the magnetization can be calculated without big effort if the Meissner effect is ideally realized, i. e. the magnetic field is completely displaced from the material. The complete displacement is of course an idealization, but we can very easily calculate further magnetic quantities, such as the magnetic susceptibility, which can be determined experimentally.

We consider a superconducting material that is placed in a homogeneous external magnetic field $\vec{H}$. Perfect realization of the Meissner effect means that the magnetic induction $\vec{B}$ is zero in the whole material volume. From the general relation $\vec{B} = \vec{H} + 4\pi\vec{M}$ between external field $\vec{H}$ and magnetization $\vec{M}$ of the material, we immediately find the relation

$$\vec{M} = -\frac{1}{4\pi}\vec{H}, \tag{1.1}$$

which determines the magnetic susceptibility in the superconducting state. This quantity describes the response of the material to an external magnetic field $\vec{H}$. Comparing (1.1) with the defining equation $\vec{M} = \chi\vec{H}$ of the magnetic susceptibility χ we find in the superconducting state the value $\chi = -1/(4\pi)$.

It turns out that in real materials the Meissner effect is not realized perfectly, i. e. there is a finite magnetic induction within a narrow region close to the surface of the superconducting material. The magnetic induction in the interior of the superconducting material is screened by surface currents flowing inside the region where the field penetrates. This area has a typical spatial extension of the order of the so-called London penetration depth λ_L (see chapter

2). Typical values of λ_L are around $500\,\text{Å}$. In the surface area, the magnetic induction $\vec{B}$ is non-zero but decays exponentially and approaches $\vec{B} = 0$ in the interior of the material (bulk superconductor) where perfect diamagnetism is found as discussed above. The typical behavior of the magnetic induction $\vec{B}$ as a function of the distance to the surface is illustrated schematically in Figure 1.4.

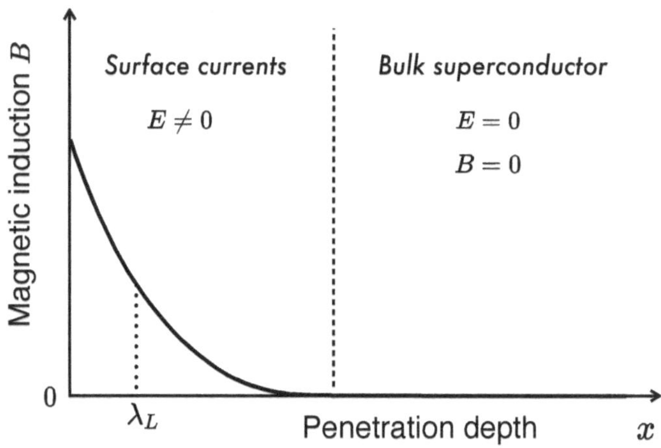

Figure 1.4: Magnetic field as a function of the distance to the surface ($x = 0$) of the superconducting material. The field decays within a typical length scale $\lambda_L \approx 500\,\text{Å}$. In this surface area, currents lead to a screening of the magnetic field inside the material.

A superconductor is an ideal conductor and therefore any finite electric field $\vec{E}$ causes an infinitely large electric current. Thus, inside a superconducting material energy conservation can only be fulfilled if the interior of the superconductor is free of any electric field, i. e. $\vec{E} = 0$. This applies, of course, to a state of thermodynamic equilibrium. Thus, Maxwell's law of induction,

$$\operatorname{rot}\vec{E} = -\frac{1}{c}\frac{\partial \vec{B}}{\partial t}$$

14

(c: speed of light in vacuum) leads for $\vec{E} = 0$ to a time-independent (static) magnetic field inside a superconductor.

1.3 – Critical magnetic field

If the superconductor is in an external magnetic field, the superconducting state initially remains stable as long as the field strength is not too large. A further increase in the field strength leads to a phase transition to the normal state. The superconducting state can either disappear completely, or a mixed state of superconducting and normal conducting domains is first formed, which is then replaced by the normal state in a second transition.

In this section, we want to collect important experimental findings on the critical values of the external magnetic field. Here the question of importance is under which conditions regarding the magnetic field the superconducting state is thermodynamically stable.

1.3.1 – Field energy

At first, let us start with a few energetic considerations. As discussed in the previous section, when applying an external magnetic field $\vec{H}$, the magnetic induction $\vec{B}$ inside the material is suppressed. Thus, there is a strong difference between $\vec{H}$ and $\vec{B}$. This occurs because the superconducting state can extremely easily induce currents on the surface whose magnetic field counteracts the external one. However, the induction of these currents and the associated dislocation of the magnetic field costs energy. If the magnetic field strength is relatively small, this energy loss is small compared to the gain in energy caused by the formation of the superconducting state. This will be shown later. Therefore, if the value of the magnetic field is smaller than some critical value, the gain in energy due to the superconducting state can be larger than the energy loss through the displacement of the field. As a result, the superconducting state is stable while the Meissner effect is present.

The amount of field energy which is 'removed' from the interior of the superconducting material can be calculated approximately as follows. If we assume that the London penetration depth λ_L is small compared to the typical dimension of the superconducting material, $\lambda_L \ll V^{1/3}$, where V is the sample volume, this energy is equivalent to the total field energy of the magnetic field inside the volume of the superconductor. In this limit it corresponds exactly to the field energy inside the space volume V if the superconductor did not exist. The related amount of field energy can be calculated from electrodynamics, $E_H = V H^2/(8\pi)$. Roughly speaking, if this energy loss E_H becomes (for particular high field strength) larger than the energy gain associated with the manifestation of the superconducting state over the normal state, the superconducting state becomes unstable.

1.3.2 – Phase transition

The above considerations lead us to the insight that at a given temperature $T < T_c$ the superconducting phase is thermodynamically stable if the external magnetic field H is lower than a certain critical magnetic field $H_c(T)$. For temperatures larger than T_c, only the metallic phase is stable. An approximate formula for $H_c(T)$, which is valid for most of the conventional superconductors, can be derived from experiments. It reads

$$H_c(T) = H_0 \left[1 - \left(\frac{T}{T_c} \right)^2 \right], \tag{1.2}$$

where H_0 is the critical field at zero temperature. The critical magnetic field curve $H_c(T)$ represents the phase boundary between the superconducting phase and the metallic phase and is qualitatively shown in Figure 1.5. Note that there are unconventional superconductors in which the phase transition line described by (1.2) is further divided into two lines. These so-called type 2 superconductors will be treated later.

Generally, one distinguishes different orders of phase transitions. A simple way to specify the order of the phase transition is to investigate the latent heat which is exchanged during this transiti

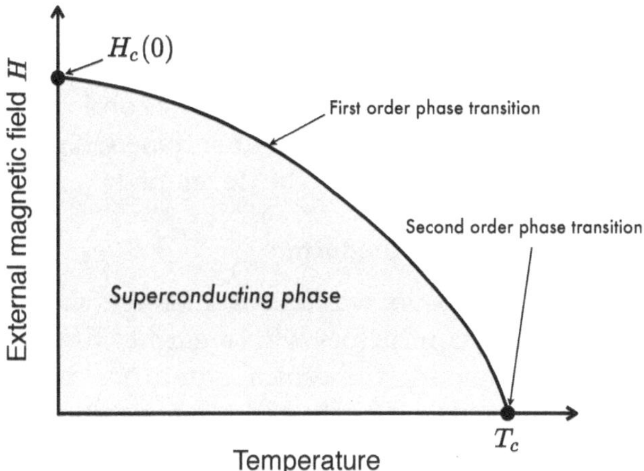

Figure 1.5: Schematic phase diagram of a conventional superconductor in the presence of an external magnetic field H. Except for $T = T_c$ and $T = 0$ the phase transition is of first order, i. e. a latent heat exchange is associated with the phase transition from the superconducting to the normal state.

on. If a finite amount of latent heat is measured, this indicates a first-order phase transition. In the case of a second-order phase transition, there would be no exchange of latent heat. A typical example of a first-order phase transition is the melting of water ice, where latent heat is supplied during the melting process.

In superconductors actual measurements reveal that for all transitions that take place at finite temperatures $T > 0$, which are less than the T_c of the field-free case, a latent heat is exchanged. The exchanged heat decreases more and more if the temperature, at which the transition takes place, is increased. Thus, along the black curve in the phase diagram in Figure 1.5 the phase transition is of first order. If the transition finally takes place exactly at T_c, i. e. without the presence of a magnetic field ($H = 0$), no latent heat is exchanged. This corresponds to a second-order phase transition. It is marked in Figure 1.5 with a point on the right edge of the phase boundary line. It can be shown theoretically that the phase transition at $T = 0$ on the left edge of the phase boundary line also corresponds to a second-order transition.

In summary, it can be said that in a finite magnetic field $H > 0$ and at finite temperature $T > 0$ the phase transition is of first order. In a field-free situation or at $T = 0$, it is of second order. We will describe and understand all the above-mentioned experimental results with the help of the theories to be developed.

1.3.3 – Type 1 and type 2 superconductors

As already mentioned above, we have to distinguish between two different types of superconductors with regard to the phase transition. This is closely linked to the typical length scales that occur in the variation of the particle density and the magnetic induction. In addition to the London penetration depth λ_L as the first important length scale, there is another one determining the properties of superconductors, the coherence length ξ. This quantity will be defined correctly in chapter 4. Here it should be only mentioned that ξ describes the characteristic length scale in which the superconducting wave function changes significantly. In addition to the London penetration depth, the length ξ can be used to further classify superconductors.

An important dimensionless parameter in this context is the *Ginzburg-Landau parameter*, defined by the ratio $\kappa := \lambda_L/\xi$. Depending on the value of κ, superconductors behave differently during the phase transition. This property is discussed in detail in chapter 5. In the following, a summary of the main results of this discussion is given.

In materials with $\kappa < 1/\sqrt{2}$, the transition takes place as described in the previous subsection: There is a critical magnetic field H_c at which the superconductivity breaks down. During the transition, the entire superconducting density disappears at the same time. The phase transition is of first order and accompanied by a latent heat transfer. Materials with $\kappa < 1/\sqrt{2}$ are called type 1 superconductors. In particular, materials with $\lambda_L \approx 0$ in which the Meissner effect is realized perfectly, belong to this family.

Materials with $\kappa > 1/\sqrt{2}$ are called type 2 superconductors. They are characterized by the existence of two critical field values, denoted by H_{c1} and H_{c2}. For external fields $H < H_{c1}$, the material behaves like a type 1 superconductor in the Meissner phase. For fields H larger than the so-called lower critical field H_{c1} the material enters an inhomogeneous superconducting phase. Here, the material has the property that a coexistence of superconducting and normal conducting regions in the material are observed. The material forms normal conducting and superconducting domains which coexist. The magnetic field partially penetrates into the material by flooding the normal conducting areas while it is still displaced from the superconducting areas.

In this phase, the magnetic flux passing through the normal conducting domains cannot be arbitrary, i. e. the field penetrates the material in a special way. It forms quantized tubes of magnetic flux which are formed by vortices of current (Abrikosov vortices). In this state, the current circulates around the normal conducting core of the vortex. Such structures were first described theoretically by Abrikosov in 1957 (Nobel Prize 2003).

The size of the Abrikosov vortex is described by the two length scales ξ and λ_L. While the core of the vortex has a characteristic size of the order of the coherence length ξ the currents flowing around the core decay on a distance of about λ_L. The circulating currents induce magnetic fields with the total flux equal to the flux quantum

$$\phi_0 = \frac{hc}{2e} = 2.07 \cdot 10^{-7}\,\mathrm{Gcm}^2, \tag{1.3}$$

where h is the Planck's constant, c the velocity of light, and e the elementary charge. Here,

$$\phi_0 = \int_{\mathscr{S}} \vec{B}\,\mathrm{d}\vec{A}$$

is the total magnetic flux of the circular current of one vortex, i. e. the surface $\mathscr{S}$ is chosen such that it encloses completely the current belonging to the particular vortex.

At H-values slightly above the lower critical field H_{c1}, the vortices arrange rather disordered. The vortex density increases with increasing magnetic field. An appropriate vortex model and the flux quantization is treated in chapter 5.

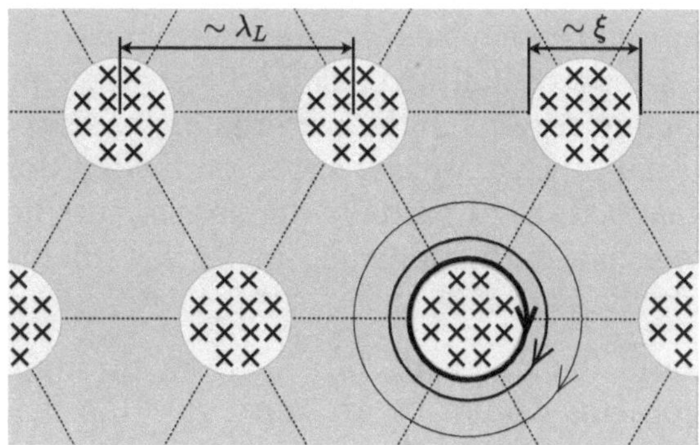

Figure 1.6: Schematic picture of the Abrikosov flux lattice in type 2 superconductors. An external magnetic field in the range $H_{c1} < H < H_{c2}$ penetrates the superconducting material (dark gray) in form of vortices (light gray). In the vortex the total magnetic flux is $\phi_0 = hc/(2e)$. The strength of the circular currents (indicated in the figure by the line thickness) decays for each vortex within a characteristic length of the order of the London penetration depth λ_L. This length scale determines also the distance between the vortex cores. For H-values close to H_{c2}, the vortices are arranged in form of a triangular lattice. The magnetic induction $\vec{B}$, schematically denoted by black crosses, points perpendicular to the plane of projection.

In a clean superconductor and for magnetic fields close to the upper critical field, the Abrikosov vortices arrange in form of a triangular lattice as illustrated in Figure 1.6. Each of the vortices carries one flux quantum. Note that, as with other lattices, defects may form as dislocations of the triangular Abrikosov lattice.

If the external magnetic field exceeds the upper critical field H_{c2}, the whole material volume enters the normal state. Typical values for the critical fields H_{c1} and H_{c2} are relatively large. For example, in NbSn$_3$ ($T_c = 18$ K) experiments find

$$H_{c1} \approx 200 \text{ G and } H_{c2} \approx 100000 \text{ G.}$$

Note that the magnetic field of the earth close to the earth's surface is approximately 0.4 G.

1.4 – Energy gap

Superconducting materials show an interesting property in the specific heat capacity. This quantity describes the ratio of the heat added to (or removed from) an object to the resulting temperature change divided by its mass. Experiments measuring the specific heat capacity as a function of the temperature show that in superconductors this quantity makes a typical jump at the critical temperature. The qualitative behavior is shown in Figure 1.7. Below we will conclude from thermodynamic arguments that such a jump has to do with the transition into an ordered phase in which latent heat is exchanged. This experimental result thus indicates that the superconducting phase has a higher order than the metallic phase.

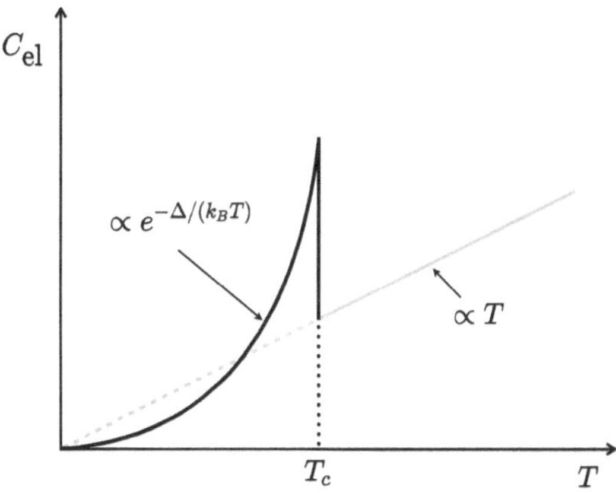

Figure 1.7: Qualitative temperature behavior of the electronic part of the specific heat for conventional superconducting materials. At the critical temperature a characteristic jump is observed.

The specific heat capacity of a solid is generally composed of different contributions. For example, thermal oscillations of the lattice ions and also the system of the conduction electrons each provide a contribution. Assuming that the superconductivity is caused by the conduction electrons, it makes sense to look at their contribution to the heat capacity.

The typical temperature behavior of the electronic specific heat $C_{el}(T)$ in a superconductor is shown in Figure 1.7. As is well-known, in the normal (metallic) phase the specific heat depends linearly on T. Instead, in the superconducting phase the material behaves differently. Experiments observe an exponential T-dependence of the form $C_{el} \propto e^{-\Delta/(k_B T)}$, where Δ is an energy constant of the order of the thermic energy $k_B T_c$ (k_B: Boltzmann constant) associated with the critical temperature T_c. The qualitatively different temperature behavior in the superconducting and normal state (black and gray lines in Figure 1.7) results in a characteristic jump of $C_{el}(T)$ at T_c.

The energy Δ is closely related to an energy gap in the spectrum of states of conduction electrons. It is also connected to the binding energy of pairs of conduction electrons (Cooper pairs) that are formed below T_c. A further indication for the existence of paired electrons in the superconducting state are the results of tunneling experiments.

Figure 1.8(a) shows a typical set up of such experiments. The superconducting material (SC) is coupled via a tunneling barrier to a normal conducting (NC) material. A bias voltage U is applied between SC and NC, which allows the charge carriers to tunnel between the two systems. The resulting tunneling current I as a function of U is plotted qualitatively in Figure 1.8(b). Usually, such an experiment leads to the result that in the superconducting state tunneling sets in only after application of a sufficiently high bias voltage $|U| > \Delta/e$, where e is the elementary charge. Here, Δ is the same energy value as can be extracted from the specific heat measurements which are described above. This observation means that for $|U| < \Delta/e$ there are no states to tunnel in. It provides a direct evidence for the existence of an energy gap in the electronic spectrum in the superconducting state.

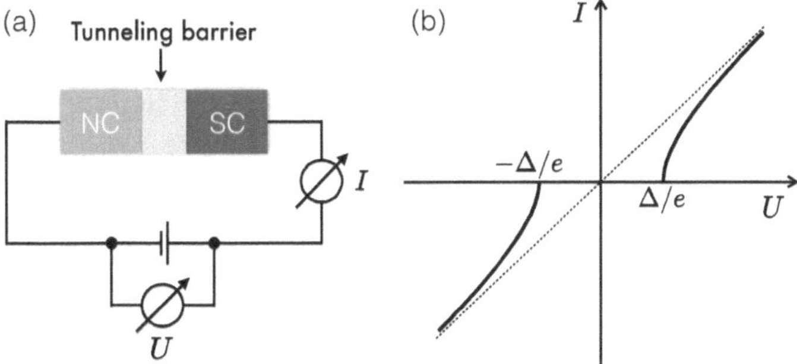

Figure 1.8: (a) Tunneling experiment to investigate the coupled arrangement of a superconductor (SC) and a normal conductor (NC) separated by a tunneling barrier. Charge carriers tunnel between SC and NC if a bias voltage U is applied. The corresponding current I as a function of U is measured. (b) Tunneling current vs. bias voltage. Tunneling sets in only after application of a sufficiently high bias voltage $|U| > \Delta/e$.

1.5 – Isotope effect

Experiments on conventional superconductors have found a close relation between the superconducting transition temperature T_c and the atomic mass M of the material. This phenomenon is called isotope effect since the influence on T_c has been discovered by considering different isotopes of the ions in the metallic material. The approximate relation found by experiments is $T_c \propto 1/\sqrt{M}$.

Thus, lattice vibrations (phonons) should play an important role for the description of conventional superconductors. The isotope effect has provided the main motivation for the development of the microscopic theory of superconductivity by Bardeen, Cooper, and Schrieffer (BCS).

1.6 – Superconducting loops

Another property of superconductors concerns closed superconducting wires (loops) in external magnetic fields described by the magnetic induction $\vec{B}(\vec{r})$, where $\vec{r}$ is the position vector. The situation is shown in Figure 1.9. There is an interesting property of such a system regarding the magnetic flux passing through any surface $\mathcal{S}$ which is enclosed by the loop. Experiments observe that the magnetic flux is always an integer multiple of a particular fixed value ϕ_0, the so-called flux quantum which is given by equation (1.3).

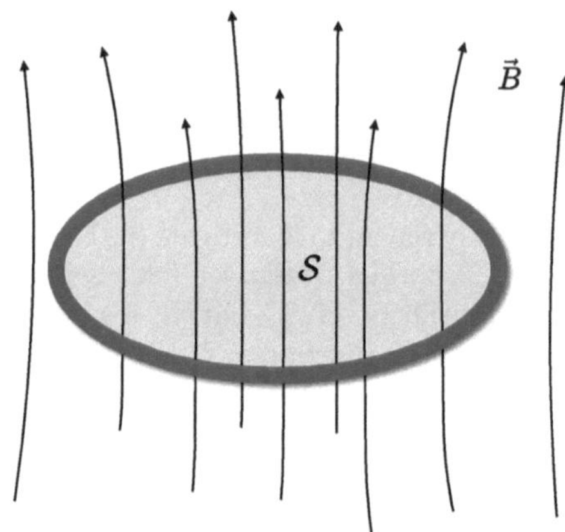

Figure 1.9: Superconducting loop (dark gray) in an external magnetic field. The magnetic flux ϕ through any surface enclosed by the loop is quantized.

Mathematically, the flux quantization is formulated as

$$\phi = \int_{\mathcal{S}} \vec{B}\, \mathrm{d}\vec{f} = \hat{n}\phi_0, \tag{1.4}$$

where $\hat{n}$ is an integer number and $\mathrm{d}\vec{f}$ is the infinitesimal surface element. Equation (1.4) means that the magnetic flux through a superconducting loop (or alternatively a hole in a bulk superconductor) cannot take an arbitrary value. Instead, loop currents are induced by the external magnetic field, and these currents modify the magnetic field inside the loop such that (1.4) is fulfilled. The so-called flux quantum $\phi_0 = hc/(2e)$ is a fixed value which can be expressed in terms of the Planck's constant, the velocity of light, and the elementary charge. The explanation of the flux quantization is given in chapter 5.

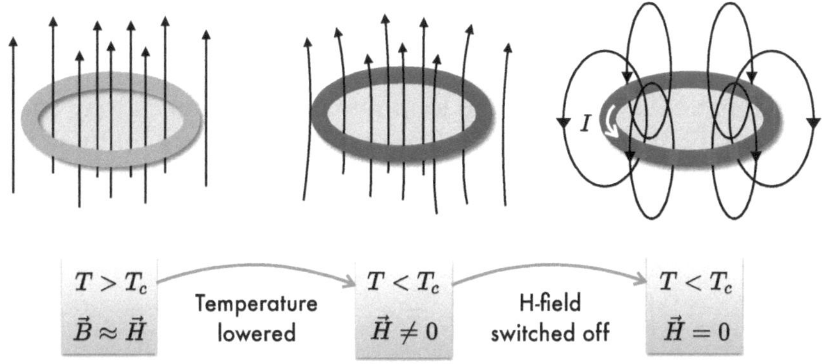

Figure 1.10: Two-step process to induce steady currents in a superconducting loop. Starting in the normal state at $T > T_c$ in an external field, the temperature is at first cooled down below T_c such that the loop becomes superconducting. In a second step the external magnetic field is switched off. Due to time-independent magnetic flux in the superconducting state, a steady current flows.

Using external magnetic fields, it is possible to induce steady currents in superconducting loops. The procedure consists of two steps and is illustrated in Figure 1.10. The procedure starts with the *normal state* of the loop, i. e. with a certain temperature T above the transition temperature T_c. Then, the loop is brought in an external field. For simplicity, let us neglect any magnetization effects of the loop, i. e. in the initial state the magnetic induction is approximately equal to the external magnetic field, $\vec{B} \approx \vec{H}$.

In a first step the material is cooled down to a temperature below T_c, thereby holding the external field $\vec{H}$ at its initial value. The material enters the superconducting state and produces a $\vec{B}$-field such that, according to (1.4), its magnetic flux ϕ has some quantized value $\phi = \hat{n}\phi_0$. Now the $\vec{B}$-field may differ somewhat from the $\vec{H}$-field.

In a second step the external magnetic field is switched off, while the loop is kept in the superconducting state. Due to Maxwell's induction law,

$$\vec{\nabla} \times \vec{E} = -\frac{1}{c}\frac{\partial \vec{B}}{\partial t},$$

the magnetic flux through the loop remains constant as long as the superconducting state is kept. This can be shown as follows. Forming the time derivative of the flux through a time-independent surface $\mathcal{S}$ enclosed by a closed curve $\mathscr{C}(\mathcal{S})$ inside the superconducting loop, we obtain

$$\frac{1}{c}\frac{d\phi}{dt} = \frac{1}{c}\int_{\mathcal{S}} \frac{\partial \vec{B}}{\partial t}\, d\vec{f} = -\int_{\mathcal{S}} \vec{\nabla} \times \vec{E}\, d\vec{f} = -\oint_{\mathscr{C}(\mathcal{S})} \vec{E}\, d\vec{r},$$

where in the last step the Stoke's theorem was used. Since the electric field is integrated along a curve inside the superconductor (where $\vec{E} = 0$) we conclude that the magnetic flux through a surface with a boundary inside the superconducting material is always constant in time. According to (1.11) the magnetic flux takes the quantized value $\phi = \hat{n}\phi_0$. Consequently, the magnetic flux passing through the loop is 'frozen', regardless of any change of the external magnetic field. The only possibility for the superconductor to keep the flux constantly flowing is to induce a magnetic field by a certain loop current I. Thereby, the current is carried by superconducting particles (supercurrent) and is characterized by a permanent flow. The term supercurrent is introduced in chapter 4 in a more general context.

Chapter 2 – London Theory

In this chapter the first attempt to explain the Meissner effect in a superconducting material is described. The London theory formulates the Maxwell equations for a material that has an infinitely large electrical conductivity. Such a material is by definition a superconductor. Since the property of superconductivity is assumed from the beginning without knowing its origin, the London theory is a phenomenological theory. In particular, effects are derived from certain assumptions about the superconducting state without providing a description how the superconducting state itself arises from the properties of the microscopic system. As we will see, we can still explain many phenomena with the theory. Above all, the theory explains the Meissner effect very well.

We start in section 2.1 with the derivation of the basic set of equations for the magnetic induction and the electrical current inside a superconductor. Based on the developed theory we will be able to explain the Meissner effect in section 2.2. In this context, the London penetration depth λ_L is introduced. Finally, in section 2.3 an alternative quantum mechanical derivation of the London equations is given.

The London theory was developed by Fritz and Heinz London in 1935. The theory describes the relationship between the electromagnetic fields of the superconductor and currents of a 'liquid' of superconducting particles inside the material. Based on Maxwell's equations, this ansatz leads to a theoretical description of the Meissner effect.

2.1 – London equations

Starting point is the assumption that the system of conduction electrons in the solid leads to an infinitely large electrical conductivity. It forms a *superconducting liquid*. The associated particle density n_s is assumed to be constant. At this point one should again emphasize that (within this phenomenological theory) the microscopic origin of the formation of a superconducting liquid remains open.

As will be shown later within the Ginzburg-Landau theory, the typical length scale in which the density of superconducting particles changes is the coherence length ξ which was already introduced in the previous chapter. Since in the London theory the particle density n_s is assumed to be constant, it can be argued that the London theory should be valid in the limit $\xi \to \infty$. This limit corresponds to a type 1 superconductor with an extremely small Ginzburg-Landau parameter $\kappa \to 0$.

We start with a general formulation of the equation of motion for a system of free-moving conduction electrons, whereby the state (normal state or superconductivity) has not yet been determined at first. Using the particle density n of the electrons, we can define a field describing the current density $\vec{j}(\vec{r}, t)$ via the following expression,

$$\vec{j}(\vec{r}, t) = n(-e)\vec{v}(\vec{r}, t), \tag{2.1}$$

where e is the elementary charge and $\vec{v}(\vec{r}, t)$ is the field of the velocity of the particle density. Here, $\vec{v}(\vec{r}, t)$ has to be understood as the velocity of the particular (infinitesimally small) volume element at position $\vec{r}$ and time t. Note that, while the density n is assumed to be constant, the vector quantities $\vec{v}$ and $\vec{j}$ are fields, i. e. they may depend on $\vec{r}$ and t.

Now we derive an equation of motion by forming the time-derivative of equation (2.1),

$$\frac{d}{dt}\vec{j} = n(-e)\frac{d}{dt}\vec{v}. \tag{2.2}$$

Since we assume that the current $\vec{j}$ is carried by conduction electrons, we may relate the total time derivative of velocity $\vec{v}$ to the acceleration of the conduction electrons in the volume element at position $\vec{r}$ and time t. Within a classical treatment of the conduction electrons as particles in an electromagnetic field, the acceleration of the electron is determined (according to Newton's law) by the electric force, a friction force proportional to the current density, which acts against the electric force, and a possible magnetic force (Lorentz force),

$$m\dot{\vec{v}} = -e(\vec{E} + \vec{v} \times \vec{B}) + \mu\vec{j}, \tag{2.3}$$

where m is the mass of one electron. Note that the friction force (friction coefficient μ) for the superconducting state is by definition equal to zero. With equation (2.1) we can replace $\vec{v}$ with $\vec{j}$,

$$\dot{\vec{j}} = \frac{ne^2}{m}\vec{E} - \frac{e}{m}\vec{j} \times \vec{B} - \frac{ne\mu}{m}\vec{j}. \tag{2.4}$$

This is the equation of motion for the current density, which describes the local temporal change of $\vec{j}$ with given fields $\vec{E}$ and $\vec{B}$. One can see from this differential equation that for a metal without a magnetic field, $\vec{B} = 0$, an equilibrium balance $\dot{\vec{j}} = 0$ for the current density is quickly set under the influence of an electric field. In this static equilibrium, the well-known Ohm's law $\vec{j} = \sigma\vec{E}$ with $\sigma = e/\mu$ results.

Now we look at the equation of motion (2.4) especially for the *superconducting* state. This is defined by the fact that the electrical conductivity σ is infinitely large, or that $\mu = 0$ applies (lack of friction). Thus, in the superconducting state the last term in equation (2.4) vanishes.

In addition, the following approximation is justified. The superconducting state is characterized by an extremely high mobility of the superconducting particles. Therefore, in the thermodynamic equilibrium the velocity of the superconducting particles must be extremely small. In addition, as shown below, the magnetic field inside the superconductor is suppressed, so that the magnetic induction $\vec{B}$ is also relatively small. Consequently, in a thermodynamic equilibrium both the velocity $\vec{v}_s$ of a superconducting electron and also $\vec{B}$ are small quantities and we are allowed to neglect the Lorentz term in (2.4). The 'stiffness' of the superconducting volume with respect to the Lorentz force is called London rigidity.

In the whole, we obtain the following equation from (2.4) in the superconducting state (index 's')

$$\frac{d\vec{j}_s}{dt} = \frac{n_s e^2}{m}\vec{E}. \tag{2.5}$$

This equation is called the first of London's equations. It determines the change of the superconducting current density in the case that an electrical field is present. One can see that the static case of a temporally constant current density always means $\vec{E} = 0$. Therefore, a current can flow, although there is no electric field. Ohm's law no longer applies.

With the equation (2.5), the property of superconductivity enters London's theory. From that, we next derive an equation which leads us to the magnetic field inside the superconductor. Forming the curl on both sides of (2.5) and applying Maxwell's equation

$$\operatorname{rot}\vec{E} + \frac{1}{c}\frac{\partial \vec{B}}{\partial t} = 0, \tag{2.6}$$

we obtain (after interchanging the spatial derivatives with the time derivative),

$$\frac{d}{dt}\left(\operatorname{rot}\vec{j}_s + \frac{n_s e^2}{mc}\vec{B}\right) = 0. \tag{2.7}$$

From this equation we may conclude that the expression inside the bracket must be equal to some time-independent vector field. Since (2.7) holds at any time t (also at the initial time t_0), we are allowed to set this vector field equal to zero if the superconducting volume is initially field-free and in thermodynamic equilibrium.

In the following, let us consider the initial setting $\vec{B}(t_0) = 0$ and $\vec{j}_s(t_0) = 0$. This situation is given, for example, when the superconductor is brought from a field-free space into a magnetic field or when the magnetic field is switched on. Under this condition we obtain the second of London's equations,

$$\text{rot}\,\vec{j}_s = -\frac{n_s e^2}{m c}\,\vec{B}\,. \tag{2.8}$$

It describes the relation between the magnetic induction inside the superconducting material and the corresponding electrical current density of superconducting particles.

2.2 – Explanation of the Meissner effect

The perhaps most prominent effort of the London theory is the explanation of the Meissner effect. For a simplified treatment, let us consider a static situation, i. e. all fields are time-independent. At first, from the first of London's equations (2.5) we can conclude that $\vec{E}(\vec{r}) = 0$ holds. In a static situation the interior of the superconductor is free of electrical fields.

To explain the Meissner effect, it is now important to consider the magnetic field $\vec{B}(\vec{r})$ inside the superconductor. For this purpose, we derive in the following a set of two equations that determine both the magnetic field $\vec{B}(\vec{r})$ and also the current density $\vec{j}_s(\vec{r})$ inside the superconductor. The first of these two equations is already given by (2.8). A second equation is found from the Ampere's law

$$\text{rot}\,\vec{B} = \vec{\nabla} \times \vec{B} = \frac{4\pi}{c}\,\vec{j}_s\,. \tag{2.9}$$

Here, the differential operator $\vec{\nabla}$ denotes the Nabla operator

$$\vec{\nabla} = \frac{\partial}{\partial x}\vec{e}_x + \frac{\partial}{\partial y}\vec{e}_y + \frac{\partial}{\partial z}\vec{e}_z,$$

where $\vec{e}_x, \vec{e}_y, \vec{e}_z$ are the unit vectors in Cartesian coordinates.

In summary, we obtain the following set of linear differential equations which determine the two fields $\vec{B}(\vec{r})$ and $\vec{j}_s(\vec{r})$ inside a superconductor,

$$\vec{\nabla} \times \vec{B} = \frac{4\pi}{c}\,\vec{j}_s, \tag{2.10}$$

$$\vec{\nabla} \times \vec{j}_s = -\frac{n_s e^2}{mc} \vec{B}\,. \tag{2.11}$$

Next, let us solve this system of equations. The first step is to form the curl on both sides of equation (2.10),

$$\vec{\nabla} \times (\vec{\nabla} \times \vec{B}) = \frac{4\pi}{c} \vec{\nabla} \times \vec{j}_s\,.$$

Replacing the right hand side with (2.11) we find

$$\vec{\nabla} \times (\vec{\nabla} \times \vec{B}) = -\frac{4\pi n_s e^2}{mc^2} \vec{B}\,. \tag{2.12}$$

The double cross product on the left hand side can be simplified using the general identity

$$\vec{\nabla} \times (\vec{\nabla} \times \vec{B}) = \vec{\nabla} \cdot (\vec{\nabla} \cdot \vec{B}) - \vec{\nabla}^2 \vec{B}$$

and Maxwell's equation $\vec{\nabla} \cdot \vec{B} = 0$. Thus, equation (2.12) leads to the following differential equation for the magnetic induction in a superconductor,

$$\vec{\nabla}^2 \vec{B} = \frac{1}{\lambda_L^2} \vec{B} \quad \text{with} \quad \lambda_L = \sqrt{\frac{mc^2}{4\pi n_s e^2}}\,. \tag{2.13}$$

The length λ_L is called London penetration depth. It has been introduced already in section 1.2 as the characteristic length scale in which the magnetic induction decays as a function of the distance to the surface of a superconductor.

To show that (2.13) describes the decay of the field, let us solve this equation by considering a superconducting material in the presence of a plain surface. We chose a coordinate system such that the surface plane coincides with the y-z plane. In this case, the surface is described by the equation $x = 0$. The magnetic induction along the surface (boundary condition) is considered to be directed parallel to the surface and in y-direction, $\vec{B}(x = 0) = B_0\,\vec{e}_y$. For simplicity, we assume that the material is isotropic with respect to y and z such that the problem reduces to one dimension and the field depends solely on x. In this case, the differential equation (2.13) simplifies to

$$\frac{\mathrm{d}^2}{\mathrm{d}x^2}\vec{B}(x) = \frac{1}{\lambda_L^2}\vec{B}(x).$$

Its general solution reads

$$\vec{B}(x) = \vec{B}(x=0)\, e^{\pm\frac{x}{\lambda_L}},$$

where $\vec{B}(x=0) = B_0\vec{e}_y$ is determined by the boundary condition. The physical relevant solution is the one with a finite field value in the limit $x \to \infty$ which is given by the solution

$$\vec{B}(x) = B_0 e^{-\frac{x}{\lambda_L}}\vec{e}_y. \tag{2.14}$$

It describes the behavior of the field as is measured by experiments. The result is shown in Figure 2.1. We can see that the field decays exponentially down to zero within a characteristic length scale given by the London penetration depth λ_L.

The superconducting current density $\vec{j}_s(x)$ can be easily calculated from Ampere's law (2.10) by forming the curl of (2.14). We obtain

$$\vec{j}_s = \frac{c}{4\pi}\vec{\nabla}\times\vec{B} = \frac{c}{4\pi}\frac{\partial B_y}{\partial x}\vec{e}_z = -\frac{c}{4\pi\lambda_L}e^{-\frac{x}{\lambda_L}}\vec{e}_z. \tag{2.15}$$

Thus, the superconducting current density decreases exponentially (within λ_L) as the $\vec{B}$-field, but points perpendicular to the $\vec{B}$-vector and parallel to the surface of the superconductor.

The solutions (2.14) and (2.15) provide a full explanation of the Meissner effect and can be interpreted as follows: An external magnetic field induces superconducting currents flowing near the surface of the superconductor. These currents induce an own magnetic field which compensates the external field in the interior of the superconductor. On the other hand, the field outside the superconductor is strengthened. This becomes clear by applying the right-hand rule to the induced magnetic field of a current flowing in

$(-\vec{e}_z)$-direction (compare Figure 2.1). These effects altogether lead to the typical displacement of the magnetic field from the interior of the superconducting material.

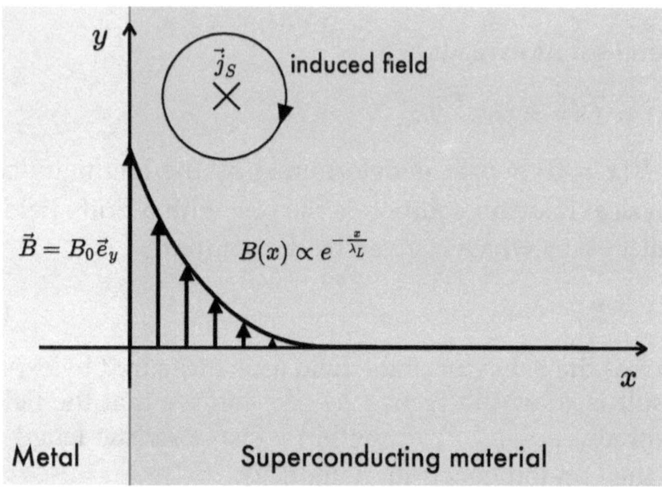

Figure 2.1: Magnetic induction (arrows and solid line) inside a superconducting material (shaded in dark gray) according to the solution of the London equations. The magnetic induction decreases to zero within the London penetration depth λ_L. This behavior is a result of a superposition of the external field and the field induced by the superconducting current $\vec{j}_s = -j_s\vec{e}_z$. The direction perpendicular to the plane of projection is represented by a cross. The direction of the induced field (circular line) is determined by the right-hand rule.

Finally, one should notice that the second of London's equations (2.8) can be written in a compact form in terms of a vector potential $\overrightarrow{A}$. The vector potential is an auxiliary quantity which is defined by its relation to the magnetic induction, $\overrightarrow{B} = \overrightarrow{\nabla} \times \overrightarrow{A}$. Due to the cross product in the definition, the vector potential is fixed up to the gradient of an arbitrary scalar field $\chi(\vec{r}, t)$, i. e.

$$\overrightarrow{A}' = \overrightarrow{A} + \mathrm{grad}(\chi)$$

is a vector potential as well (gauge invariance). This property is used in electrodynamics to bring Maxwell's equations in a specific form adapted to a particular problem. A possible vector potential inside a superconductor is given by

$$\vec{A} = -\frac{mc}{n_s e^2}\,\vec{j}_s\,.$$

(2.16)

This relation is easily seen by forming the curl on both sides, which leads with $\vec{B} = \vec{\nabla} \times \vec{A}$ immediately to (2.8).

According to the continuity equation, our initial assumption of a constant superconducting particle density n_s is equivalent to

$$\vec{\nabla} \cdot \vec{j}_s = 0.$$

Thus, the vector potential has the property

$$\vec{\nabla} \cdot \vec{A} = 0.$$

(2.17)

This equation fixes the vector potential, and it can be considered as a particular gauge (London's gauge).

2.3 – Quantum mechanical derivation

At the end of this chapter, we want to give an alternative derivation of the London equations, which is based on a quantum theoretical formulation. The aim is to show that the London theory is compatible with the quantum theory. Readers who are not familiar with the concepts of quantum theory can skip this section, because it is not absolutely necessary for the understanding of later chapters.

We start with the general quantum mechanical form of the electric current density for a system of N_e conduction electrons in a magnetic field. The operator of the current density reads

$$\vec{\mathscr{J}}(\vec{r}) = \frac{-e}{2m}\sum_{i=1}^{N_e}\left\{ \left[\vec{\mathscr{P}}_i + \frac{e}{c}\vec{A}(\vec{\mathscr{R}}_i)\right]\delta(\vec{r} - \vec{\mathscr{R}}_i) \right.$$
$$\left. + \delta(\vec{r} - \vec{\mathscr{R}}_i)\left[\vec{\mathscr{P}}_i + \frac{e}{c}\vec{A}(\vec{\mathscr{R}}_i)\right]\right\},$$

(2.18)

where $\vec{\mathcal{R}}_i$ and $\vec{\mathcal{P}}_i$ are the position and momentum operators of one electron indexed by number i, respectively. Note that the second term in (2.18) is the Hermitian conjugate of the first term. The magnetic field is described by a vector potential $\vec{A}(\vec{r})$. In the following, we always mark quantum mechanical operators with calligraphic letters.

To show that the general form (2.18) of the current density is consistent with our notation of the superconducting current density $\vec{j}_s$, it is advantageous to decompose the current operator into two parts,

$$\vec{\mathcal{J}}(\vec{r}) = \vec{\mathcal{J}}_{\text{para}}(\vec{r}) + \vec{\mathcal{J}}_{\text{dia}}(\vec{r}),$$

where

$$\vec{\mathcal{J}}_{\text{para}}(\vec{r}) = \frac{-e}{2m} \sum_{i=1}^{N_e} \left[\vec{\mathcal{P}}_i \delta(\vec{r} - \vec{\mathcal{R}}_i) + \delta(\vec{r} - \vec{\mathcal{R}}_i) \vec{\mathcal{P}}_i \right], \quad (2.19)$$

is the paramagnetic part, and

$$\vec{\mathcal{J}}_{\text{dia}}(\vec{r}) = \frac{-e^2}{mc} \mathcal{N}_s(\vec{r}) \vec{A}(\vec{r}), \qquad (2.20)$$

with

$$\mathcal{N}_s(\vec{r}) = \sum_{i=1}^{N_e} \delta(\vec{r} - \vec{\mathcal{R}}_i),$$

can be considered as the diamagnetic part of the current density. From the beginning we have assumed that the superconducting liquid is carried by the conduction electrons. Therefore, the operator $\mathcal{N}_s(\vec{r})$ in (2.20) describes the operator of the superconducting particle density and the previously introduced observable n_s can be considered as the expectation value of $\mathcal{N}_s$, i. e. $n_s = \langle \mathcal{N}_s \rangle$.

The total current $\vec{\mathcal{J}}$ is made up of a paramagnetic part $\vec{\mathcal{J}}_{\text{para}}$ and a diamagnetic part $\vec{\mathcal{J}}_{\text{dia}}$. The paramagnetic part describes the current due to the proper motion of the conduction electrons which is represented by the momentum operators $\vec{\mathcal{P}}_i$. The diamagnetic part emerges through the coupling between density and vector potential. It describes the motion due to the presence of a magnetic field.

The aim is to derive the second London equation in the form (2.16). For this purpose, we consider the current density of the superconducting particles as a measurable observable instead of an operator. In quantum theory, the relationship between operator and observable is described by the expectation value. In the following, we calculate the expectation value of the current density and show that the paramagnetic part of the current density disappears even if a magnetic field is present.

Let us start with the field-free case $\vec{A} = 0$, characterized by $\vec{\mathcal{J}}_{\text{dia}} = 0$. We assume that the system is in its *ground state* $|G\rangle$ which is exactly taken by the system at zero temperature, $T = 0$. In this case, the superconducting current is carried by its paramagnetic part. The corresponding expectation value is given by

$$\vec{j}_s(\vec{r}, \vec{B} = 0) = \langle G | \vec{\mathcal{J}}(\vec{r}) | G \rangle = \langle G | \vec{\mathcal{J}}_{\text{para}}(\vec{r}) | G \rangle.$$

However, in the thermodynamic equilibrium of a field-free superconductor, the average value of the paramagnetic current vanishes as well. Thus, for $\vec{B} = 0$ and $T = 0$ (when the system takes its ground state), there is no measurable current, i. e.

$$\vec{j}_s(\vec{r}, \vec{B} = 0) = 0.$$

London's basic assumption is now that in the presence of a magnetic field $\vec{A} \neq 0$ the superconducting ground state is not changed. This is the quantum mechanical formulation of the London rigidity (compare section 2.1). Since the operator $\vec{\mathcal{J}}_{\text{para}}$ is independent on $\vec{A}$, we may argue that (in thermodynamic equilibrium) the paramagnetic part of the current remains zero, i. e.,

$$\langle G | \vec{\mathcal{J}}_{\text{para}}(\vec{r}) | G \rangle = 0$$

holds also for $\vec{A} \neq 0$. Thus, in the presence of a magnetic field, the superconducting current is solely determined by the diamagnetic part,

$$\vec{j}_s(\vec{r}, \vec{A}) = \langle G | \vec{\mathcal{J}}(\vec{r}) | G \rangle = \langle G | \vec{\mathcal{J}}_{\mathrm{dia}}(\vec{r}) | G \rangle$$
$$= \frac{-e^2}{mc} \langle G | \mathcal{N}_s(\vec{r}) | G \rangle \, \vec{A}(\vec{r}), \tag{2.21}$$

where in the last step the expression of $\vec{\mathcal{J}}_{\mathrm{dia}}(\vec{r})$ from (2.20) was used. The matrix element $\langle G | \mathcal{N}_s(\vec{r}) | G \rangle$ is the expectation value of the superconducting particle density operator with respect to the ground state. We have argued that the system stays in its ground state also in the presence of a magnetic field. Therefore, the property that the expectation value in (2.21) represents the density n_s of superconducting particles holds also for $\vec{A} \neq 0$. This leads us directly to the result

$$\vec{j}_s(\vec{A}) = -\frac{n_s e^2}{mc} \vec{A}, \tag{2.22}$$

which exactly corresponds to the second of London's equations in the formulation (2.16) in terms of the vector potential.

2.4 – Concluding remarks

Let us now conclude this chapter with some final remarks and discussion. This chapter has presented a phenomenological theory to explain properties of superconductors (for example the Meissner effect) if a superconducting 'liquid' of conduction electrons is assumed from the beginning. Thereby, the microscopic origin of the superconducting condensate remains open. The central property of superconductivity enters in the first of London's equations (2.5). The Maxwell's equations and the second of London's equations (2.8) describe the magnetic field and superconducting current inside a superconductor in thermodynamic equilibrium.

The London theory is based on two assumptions. Firstly, the superconducting particle density is assumed to be constant. Secondly, excitations of higher energy states are neglected. These two approximations are valid for superconductors with a large coherence length ξ in comparison to the London penetration depth λ_L, i. e. $\xi \gg \lambda_L$. From the second assumption follows that the London theory is only valid for very low temperatures in comparison to the transition temperature, $T \ll T_c$. In most of the conventional superconductors experiments find a relatively large coherence length. Consequently, the London theory is applicable for conventional superconductors which are typical type 1 superconductors with small Ginzburg-Landau parameters, $\kappa \ll 1$.

There are attempts to generalize the London theory to make it applicable also to superconductors with higher values of κ. One example is the theory by Pippard which generalizes formula (2.22) to a non-local form by assuming that the vector potential decays within a characteristic coherence length. This effect is similar to the mean free path in normal metals and leads to a weaker superconducting current. The important gain of Pippard's generalization is an explanation of the phenomenon that in dirty superconductors the London penetration depth is observed to be larger than in clean ones. A phenomenological theory valid for superconductors with higher κ values and temperatures close to T_c is presented in chapter 4. For this, however, we first need some basic terms and definitions from thermodynamics. These will be covered in the following chapter.

Chapter 3 – Thermodynamic basics

The following chapter provides an introduction to the basic concepts of the thermodynamics which is needed for the later introduction of the Ginzburg-Landau theory. We start in the section 3.1 with the main theorems of thermodynamics. Then in the next section 3.2 we find the relationship between the internal energy and the electromagnetic energy and introduce the concept of the thermodynamic potential. In section 3.3 we introduce the basic concepts of the functional and its derivative. Using these considerations we introduce in the sections 3.4 and 3.5 important thermodynamic potentials which are needed for the description of the superconducting transition. Finally, in section 3.6 we give an introduction to Landau's general theory of phase transitions. Thereby, concepts of the Landau potential and order parameters are introduced and discussed in the context of the superconducting transition.

3.1 – Laws of Thermodynamics

A phenomenological theory of superconductivity describes the state of a macroscopic system based on general thermodynamic arguments. Thus, a reasonable starting point is to apply the laws of thermodynamics to a material which can take a superconducting state and which is in an electromagnetic field. The first law of thermodynamics states that in the case of fixed particle number a change in the internal energy dU only occurs when the system is exposed to the following two possible processes: (i) The system is supplied with an infinitesimal amount of heat δQ. (ii) The electromagnetic energy contained in the material changes as a result of a change in the electric and magnetic fields. Note that volume work does not matter, as no pressure is exerted on the material. Thus, up to now the first law of thermodynamics reads $dU = \delta Q + \delta E_m$, where δE_m is an infinitesimal variation of the electromagnetic energy contained in the material.

3.2 – Heat and electromagnetic energy

Now we express the energy quantities δQ and δE_m by infinitesimal variations of state variables. According to the second law of thermodynamics, the amount of heat δQ, which is supplied at a temperature T of the system, is accompanied by an infinitesimal change dS of the entropy S of the system. The relation between the heat and the corresponding change of entropy is $\delta Q = T dS$.

Next we turn to the electromagnetic energy δE_m. The change in the electromagnetic energy contained in a volume V can easily be calculated from the law of conservation of energy in electrodynamics of matter,

$$\operatorname{div}\vec{S}_m = -\frac{1}{4\pi}\left(\vec{H}\cdot\frac{\partial\vec{B}}{\partial t} + \vec{E}\cdot\frac{\partial\vec{D}}{\partial t}\right) - \vec{E}\cdot\vec{j}. \qquad (3.1)$$

The quantities contained in this equation have the following meaning: $\vec{S}_m$ is the electromagnetic energy current density, $\vec{H}$ and $\vec{E}$ are the external magnetic and electric fields, $\vec{B}$ and $\vec{D}$ are the magnetic and electric inductions, $\vec{j}$ is the electric current density.

Integrating the left side of (3.1) over the volume of the material and applying the Gauss theorem, one gets the electromagnetic energy flowing out of a surface $\mathcal{S}$ which encloses the entire volume $\mathcal{V}$ (per unit of time). This corresponds to the negative change in the electromagnetic energy contained in the volume $\mathcal{V}$, as the total energy decreases when the energy flow is directed to the outside. So we can write:

$$\int_{\mathcal{V}} \operatorname{div}\vec{S}_m \, d^3r = \int_{\mathcal{S}} \vec{S}_m\cdot d\vec{f} = -\frac{dE_m}{dt}.$$

Applying this volume integral to the right side of (3.1) leads us to

$$\frac{dE_m}{dt} = \int_{\mathcal{V}}\left[\frac{1}{4\pi}\left(\vec{H}\cdot\frac{\partial\vec{B}}{\partial t} + \vec{E}\cdot\frac{\partial\vec{D}}{\partial t}\right) + \vec{E}\cdot\vec{j}\right] d^3r. \qquad (3.2)$$

It is known from the London theory that the electric field in a superconductor can be neglected. Therefore, in the following we can omit all terms proportional to $\vec{E}$ and a change in electromagnetic energy is determined solely by changing the magnetic field. The infinitesimal change in electromagnetic energy due to an infinitesimal variation of the magnetic field can be expressed according to (3.2) as follows,

$$\delta E_m = \frac{1}{4\pi} \int_{\mathcal{V}} d^3 r \vec{H} \cdot d\vec{B},$$

where both the external magnetic field $\vec{H}$ and the corresponding variation of the magnetic induction $d\vec{B}$ generally depend on the position vector $\vec{r}$.

Finally, with the above expression for δE_m and $\delta Q = TdS$ the first law of thermodynamics in a superconductor reads

$$dU = TdS + \frac{1}{4\pi} \int_{\mathcal{V}} d^3 r \vec{H} \cdot d\vec{B}. \tag{3.3}$$

From this equation one can see that S and $\vec{B}$ are the natural variables of the internal energy, i. e. $U = U[S, \vec{B}]$ is a thermodynamic potential. Note that the internal energy must generally be considered as a *functional* with respect to the field $\vec{B}(\vec{r})$. This is indicated by the square brackets. The thermodynamic potential is used to derive the equations of state from it.

3.3 – Functional and functional derivation

At this point, it is convenient to give a brief mathematical insertion to the concept of a functional and its derivation. It is the problem that a function is dependent on a continuum of variables. How can the well-known concept of a function of several variables and the associated partial derivative be extended to a continuum of variables? We carry out these considerations on the basis of a function that has the special form of the thermodynamic potentials considered here. This function $F(\psi_0, \ldots, \psi_N)$, which depends on the N variables $\psi_0, \ldots, \psi_N$, has the following additive form,

$$F(\psi_0, \ldots, \psi_N) = \sum_{m=0}^{N} f(\psi_m), \tag{3.4}$$

where $f(x)$ is an arbitrary function. In this case, the series of the usual partial derivatives can be calculated easily,

$$\frac{\partial F}{\partial \psi_n} = f'(\psi_n) \quad \text{with} \quad n = 0, \ldots, N. \tag{3.5}$$

This series of derivatives describes the 'sensitivity' of F with respect to 'local' variations represented by the index number n.

Next, we move from the discrete to the continuous description. In this continuum limit, the set of variables ψ_n turns into a field $\psi(\vec{r})$ and the summation in (3.4) turns into an integral,

$$F[\psi(\vec{r})] = \int_{\mathscr{V}} \mathrm{d}^3 r' \, f(\psi(\vec{r}')), \tag{3.6}$$

which corresponds to an arbitrary extensive quantity in thermodynamics. The assignment of the field $\psi(\vec{r})$ to the scalar value $F[\psi(\vec{r})]$ is called *functional*. This is indicated by the square brackets. It also happens that a vector field $\vec{B}(\vec{r})$ is assigned to a scalar value by means of a functional. In this case, the functional has the form

$$F[\vec{B}(\vec{r})] = \int_{\mathscr{V}} \mathrm{d}^3 r' \sum_{i=1}^{3} f_i(B_i(\vec{r}')), \tag{3.7}$$

where the index ‚i' denotes the components of a vector.

Next, we define a local variation analog to the discrete case of the partial derivative as it is given by (3.5). The functional derivative is defined by

$$\frac{\delta F}{\delta \psi(\vec{r})} := f'(\psi(\vec{r})). \tag{3.8}$$

It represents the series of partial derivatives (3.5) in the continuum limit, where the index n is now replaced by the position vector $\vec{r}$. Note that the functional derivative is by its definition (3.8) a scalar field describing the 'local' variations with respect to $\psi(\vec{r})$ at position $\vec{r}$. If a functional of a vector field according to the above definition (3.7) is given, the functional derivative is defined analogously (3.8),

$$\frac{\delta F}{\delta B_i(\vec{r})} := f_i'(B_i(\vec{r})). \tag{3.9}$$

Note that in this case the three components i of the functional derivative together form a new vector field which sometimes is denoted as $\delta F/\delta \vec{B}$.

The concept of the functional derivative is often used in the following, since the variables of the thermodynamic potentials considered here are often fields. With the acquired knowledge, we can next introduce the free energy as another thermodynamic potential.

3.4 – Free energy

As explained in more detail below, the basic idea of the theory of phase transitions is the introduction of an additional parameter in the thermodynamic potential. In the low temperature phase, this so-called order parameter is then obtained as a function of the natural variables of the thermodynamic potential.

In experiments, the phase transition is usually described as a function of the temperature T and not as a function of entropy, because entropy is rather difficult to measure. Therefore, the thermodynamic potential that has T and $\vec{B}$ as natural variables is usually used to describe the phase transition. This is called the free energy

$$F = F[\vec{B}, T].$$

It is defined through a Legendre transformation from U as follows,

$$F := U - TS. \tag{3.10}$$

It is easily seen that F defined in this way has $\vec{B}$ and T as natural variables. We simply apply the total differential to (3.10) and use the above equation (3.3) for dU,

$$dF = dU - d(TS) = dU - SdT - TdS$$
$$= TdS + \frac{1}{4\pi} \int_{\mathcal{V}} d^3r \vec{H} \cdot d\vec{B} - SdT - TdS,$$

and we obtain,

$$dF = \frac{1}{4\pi} \int_{\mathcal{V}} d^3r \vec{H} \cdot d\vec{B} - SdT. \tag{3.11}$$

With that, it is shown that the natural variables of F are $\vec{B}$ and T.

Later we will see that the free energy in the superconducting state can be expressed more easily as a functional of the vector potential $\vec{A}$ instead of as a functional of the magnetic induction $\vec{B}$. Since $\vec{A}$ and $\vec{B}$ are related with each other via the equation $\vec{B} = \mathrm{rot}\,\vec{A}$, $\vec{A}$ can alternatively be considered as a natural variable of the free energy instead of $\vec{B}$. We can easily show this by rewriting (3.11) using the replacement $\vec{B} = \vec{\nabla} \times \vec{A}$. At first, we obtain

$$dF = \frac{1}{4\pi} \int_{\mathcal{V}} d^3r \vec{H} \cdot (\vec{\nabla} \times d\vec{A}) - SdT.$$

The vector product inside the volume integral can be replaced using the general differential operator rule

$$\vec{\nabla} \cdot (\vec{a} \times \vec{b}) = \vec{b} \cdot (\vec{\nabla} \times \vec{a}) - \vec{a} \cdot (\vec{\nabla} \times \vec{b})$$

which is valid for arbitrary vector fields $\vec{a}$ and $\vec{b}$. We find

$$dF = \frac{1}{4\pi} \int_{\mathcal{V}} d^3r \vec{\nabla} \cdot (d\vec{A} \times \vec{H})$$
$$+ \frac{1}{4\pi} \int_{\mathcal{V}} d^3r\, d\vec{A} \cdot (\vec{\nabla} \times \vec{H}) - SdT.$$

The first term can be replaced with a surface integral with the help of the Gauss theorem, which disappears if one chooses the surface in the infinite. Thus, we obtain

$$\mathrm{d}F = \frac{1}{4\pi} \int_{\mathcal{V}} \mathrm{d}^3 r\, \mathrm{d}\vec{A} \cdot (\vec{\nabla} \times \vec{H}) - S\mathrm{d}T, \tag{3.12}$$

which shows that F can alternatively be considered as a functional with respect to the natural variables $\vec{A}$ and T.

With the knowledge of the thermodynamic potential $F[\vec{B},T]$ or $F[\vec{A},T]$, two thermodynamic quantities can be explicitly calculated. At first, one can see from equation (3.11) that the negative partial derivative of F with respect to T (at fixed $\vec{B}$) always leads us to the entropy of the system,

$$S[\vec{B},T] = -\left. \frac{\partial F[\vec{B},T]}{\partial T} \right|_{\vec{B}}. \tag{3.13}$$

Furthermore, the concept of the functional derivative introduced above can also be used to represent the field term of the total differential (3.11) in the form of a derivative. We can write

$$\mathrm{d}F = \int_{\mathcal{V}} \mathrm{d}^3 r\, \left. \frac{\delta F[\vec{B},T]}{\delta \vec{B}} \right|_{T} \cdot \mathrm{d}\vec{B} + \left. \frac{\partial F[\vec{B},T]}{\partial T} \right|_{\vec{B}} \mathrm{d}T. \tag{3.14}$$

A comparison of this equation with (3.11) leads on the one hand to the already known relation (3.13) and on the other hand one can see that the functional derivative of F (at fixed T) is related to the external magnetic field as follows,

$$\vec{H}[\vec{B},T] = 4\pi \left. \frac{\delta F[\vec{B},T]}{\delta \vec{B}} \right|_{T}, \tag{3.15}$$

which describes the relation between $\vec{H}$ and $\vec{B}$. Alternatively, if the functional $F[\vec{A},T]$ is given the comparison with (3.12) leads us to

$$\vec{\nabla} \times \vec{H} = 4\pi \left. \frac{\delta F[\vec{A},T]}{\delta \vec{A}} \right|_T . \tag{3.16}$$

The relations (3.15) or (3.16) describe the property of the system to response to an external magnetic field. From electrodynamics in matter, it is known that the relationship between $\vec{H}$ and $\vec{B}$ must always have the form

$$\vec{H} = \vec{B} - 4\pi \vec{M}, \tag{3.17}$$

where $\vec{M}$ is called the *magnetization* of the system. If one compares this general relationship with (3.15), one can see that the free energy must be composed of two additive parts, $F = F_B + F_M$, from which the functional derivation of the first part must lead back to $\vec{B}$ itself. Thus, from the equations (3.15) and (3.17) we find

$$F[\vec{B},T] = F_B[\vec{B}] + F_M[\vec{B},T], \tag{3.18}$$

with

$$\frac{\delta F_B[\vec{B}]}{\delta \vec{B}} = \frac{1}{4\pi}\vec{B} \quad \text{and} \quad \left. \frac{\delta F_M[\vec{B},T]}{\delta \vec{B}} \right|_T = -\vec{M}. \tag{3.19}$$

From the first equation one can see that it is possible to find an explicit expression for the field part F_B. On the other hand, the second part F_M must first remain open, because one generally does not know the magnetization of the system.

At first, we determine explicitly the first part F_B with the help of the properties of the functional derivative introduced above. Because of (3.7) we can first write

$$F_B[\vec{B}] = \int_{\mathcal{V}} d^3r' \sum_{i=1}^{3} f_{B,i}(B_i(\vec{r}')), \tag{3.20}$$

where the $f_{B,i}(B_i)$ are energy density functions distinguishing the three components of $\vec{B}$. To determine them we use the definition of the functional derivative, equation (3.9), and the explicit result of it as it is given by the first equation of (3.19),

$$\frac{\delta F_B}{\delta B_i} = f'_{B,i}(B_i) = \frac{1}{4\pi} B_i \,.$$

Integration over B_i results in $f_{B,i}(B_i) = B_i^2/(8\pi)$ and thus we obtain from (3.20),

$$F_B[\vec{B}] = \frac{1}{8\pi} \int_{\mathscr{V}} \mathrm{d}^3 r\, \vec{B}^2(\vec{r}) \,.$$

If we use this result in the decomposition (3.18), we finally obtain the expression for the free energy, which is to be used in the following for the description of the superconducting phase transition,

$$F[\vec{B},T] = \frac{1}{8\pi} \int_{\mathscr{V}} \mathrm{d}^3 r\, \vec{B}^2(\vec{r}) + F_M[\vec{B},T] \,. \tag{3.21}$$

The task of the theory to be dealt with in the following is to find an explicit expression for the remaining part F_M, which is valid at least near the phase transition. If this expression is known, then the two equations (3.13) and (3.15) can be considered as the equations of state of the thermodynamic system.

3.5 – Gibbs free energy

In later considerations, especially in the study of the H-T phase diagram, one still needs the thermodynamic potential, which has $\vec{H}$ and T as natural variables. This quantity is called Gibbs free energy $G = G[\vec{H},T]$. It is defined as the Legendre transformation of the free energy F as follows,

$$G := F - \frac{1}{4\pi} \int_{\mathscr{V}} \mathrm{d}^3 r\, \vec{H} \cdot \vec{B} \,. \tag{3.22}$$

Again, it can be easily shown that G defined in this way has $\vec{H}$ and T as natural variables. We simply apply the total differential to (3.22) and use the equation (3.11) for $\mathrm{d}F$,

$$\begin{aligned}
\mathrm{d}G &= \mathrm{d}F - \frac{1}{4\pi}\int_{\mathcal{V}} \mathrm{d}^3 r \left(\vec{B}\cdot \mathrm{d}\vec{H} + \vec{H}\cdot \mathrm{d}\vec{B}\right)\\
&= \frac{1}{4\pi}\int_{\mathcal{V}} \mathrm{d}^3 r\,\vec{H}\cdot \mathrm{d}\vec{B} - S\mathrm{d}T\\
&\quad - \frac{1}{4\pi}\int_{\mathcal{V}} \mathrm{d}^3 r \left(\vec{B}\cdot \mathrm{d}\vec{H} + \vec{H}\cdot \mathrm{d}\vec{B}\right),
\end{aligned}$$

and we obtain,

$$\mathrm{d}G = -\frac{1}{4\pi}\int_{\mathcal{V}} \mathrm{d}^3 r\,\vec{B}\cdot \mathrm{d}\vec{H} - S\mathrm{d}T. \qquad (3.23)$$

With that, it is shown that the natural variables of G are indeed $\vec{H}$ and T.

Analogous to the above considerations on the free energy, state variables can also be calculated here from the equation (3.23). First of all, we also get the entropy here if we derive G partially with respect to T and fix the field variable (in this case $\vec{H}$),

$$S[\vec{H},T] = -\left.\frac{\partial G[\vec{H},T]}{\partial T}\right|_{\vec{H}}. \qquad (3.24)$$

Secondly, we again get a connection between $\vec{B}$ and $\vec{H}$ by functional derivation of G with respect to $\vec{H}$ at fixed temperature,

$$\vec{B}[\vec{H},T] = -4\pi\left.\frac{\delta G[\vec{H},T]}{\delta \vec{H}}\right|_{T}. \qquad (3.25)$$

Note that this equation represents the inverse map to (3.15), i. e. one can calculate $\vec{B}$ out of a given $\vec{H}$. Again it should be noted that the relationship $\vec{B} = \vec{H} + 4\pi\vec{M}$ is predetermined by electrodynamics in matter. Thus, analogous to the above considerations on the free energy, it follows that the Gibbs free energy is also composed of two parts additively. From (3.25) follows here

$$G[\vec{H}, T] = -\frac{1}{8\pi} \int_{\mathcal{V}} \mathrm{d}^3 r \vec{H}^2(\vec{r}) + G_M[\vec{H}, T], \qquad (3.25)$$

where

$$\left. \frac{\delta G_M[\vec{H}, T]}{\delta \vec{H}} \right|_T = -\vec{M}.$$

We will use these relationships later to investigate the critical magnetic field. First, however, we will examine the phase transition as a function of the variables $\vec{B}$ and T. Thus, it makes sense to first start with the free energy as the thermodynamic potential.

3.6 – Landau theory of phase transitions

Before the thermodynamic potential can be explicitly formulated for the superconducting state, we first need the general concept of distinguishing between different phases and describing the transi-

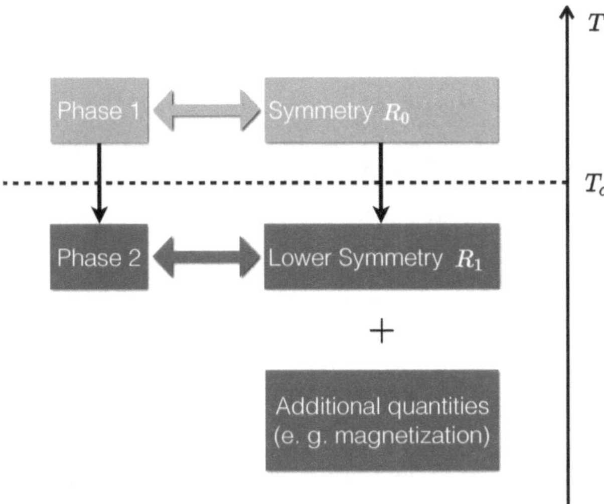

Figure 3.1: Schematic picture of a phase transition within the Landau theory.

50

tion between them. This is to be done here with the help of the Landau theory of phase transitions. The basic idea of this theory is as follows.

The properties of a solid are closely related to the *symmetry of the underlying macroscopic state* of the solid. The starting point of the theory is that a phase transition is generally combined with a change of the symmetry of the system. For example, during the transition from liquid water to water ice, a change in the translational symmetry in the spatial arrangement of the water molecules is associated. While in water ice the molecules are arranged symmetrically according to a certain rule, they do not have any specific arrangement in liquid water. This is generally associated with a change in the value of certain physical quantities during the transition. When freezing water, this is the change in volume, as the molecular arrangement in the ice produces a lower density compared to water.

If the symmetry changes during the transition from an unordered to an ordered state, at least one additional parameter is needed to describe the ordered state. In the case of water ice, this parameter can be, for example, the size of the elementary cell in the ice. In a solid, other additional quantities can occur that belong to phase transitions, such as magnetization, dielectric polarization, or in the case of a superconductor, the superconducting condensate. Each of these transitions includes a parameter that describes the respective ordered state. This parameter is called an order parameter. By definition, it is equal to zero in the disordered phase and different from zero in the ordered phase. During the transition, the state of the system can either change abruptly (first order phase transition) or the new ordered state continuously develops from the disordered state (second order phase transition).

The idea of a general phase transition is shown in Figure 3.1. Exactly at the transition point $T = T_c$, the states of the high- and low-temperature phase coincide. Consequently, the space group of one phase must be a subgroup of the other phase. In most cases (in particular in superconductors) the higher symmetry phase corresponds to the high-temperature phase and the lower symmetry phase corresponds to the low-temperature phase.

In the following, our focus lies on the phase transition from the metallic state to the superconducting state. According to the considerations of section 3.4, we assume that the system in the metallic state (normal state) can be completely described by the magnetic induction $\vec{B}$ and the temperature T. The normal state then corresponds to the high symmetry state.

If the system undergoes the phase transition by lowering the temperature below T_c, it enters a lower symmetry, and, hence, the description of the system only by $\vec{B}$ and T (as in the high symmetry state) becomes insufficient. Thus, an additional variable, which generally occurs in the form of a field, is required: the *order parameter* $\psi = \psi(\vec{r})$. It has to be included as an additional variable in the free energy, $F[\vec{B}, T] \rightarrow F[\vec{B}, T, \psi]$. Note that the magnetic induction $\vec{B}$ in the superconducting state usually differs significantly from the magnetic induction in the normal state as it is shown below.

The order parameter, which is supposed to describe the superconducting phase, is first of all a scalar field with specific properties, which we will determine in detail in the next chapter. For the description of the phase transition within the framework of the Landau theory, only the property of the *presence* of an order parameter in the superconducting state is important. This means that the order parameter for a certain constellation of the state variables T and $\vec{B}$ is either equal to zero (normal state) or is different from zero (superconducting state). The actual field values of the order parameter as a function of the position is determined from the requirement that the free energy in equilibrium becomes minimal. This method will be discussed in detail in the next chapter for various cases.

Chapter 4 – Ginzburg-Landau theory

The Ginzburg-Landau theory investigates macroscopic properties of superconductors by use of general thermodynamic arguments. It was developed by Ginzburg (Nobel Prize 2003), Landau, Abrikosov, and Gorkov. This phenomenological theory is a special form of the Landau theory of phase transitions as introduced in the previous section 3.6. The specification is that the order parameter is given a particular property, so that the ordered state has an infinitely large electrical conductivity. As we will show later, this is only possible if the order parameter is a *complex number* and the phase of the order parameter is given the freedom to vary spatially. Thus, the theory assumes from the beginning that the order parameter of the superconducting phase — in the following called superconducting order parameter — has the complex form

$$\psi(\vec{r}) = |\psi(\vec{r})|\, e^{i\Theta(\vec{r})}, \qquad (4.1)$$

where its absolute value $|\psi(\vec{r})|$ and also its phase $\Theta(\vec{r})$ generally depend on the position vector $\vec{r}$. Thus, the superconducting order parameter is characterized by the two real fields $|\psi(\vec{r})|$ and $\Theta(\vec{r})$ or, alternatively, by one complex field $\psi(\vec{r})$. The spatially varying phase appears here as an additional degree of freedom that must enter the theory to describe superconductivity. We will later show that this property allows an electric current without an external electric field. The microscopic origin of the phase of the order parameter remains unknown which means that the Ginzburg-Landau theory is a phenomenological theory.

One can also see the field $\psi(\vec{r})$ as a quantum mechanical state. Thus, it can be assumed that the superconducting particles form a *macroscopic wave function* $\psi(\vec{r})$ which describes a superconducting condensate. Therefore, it is obvious to set this wave function equal to the order parameter of the superconducting phase. As any other wave function, $\psi(\vec{r})$ can be considered in quantum mechanics as a scalar product of state vectors in the Hilbert space,

$$\psi(\vec{r}) = \langle \vec{r} \,|\, \psi \rangle. \qquad (4.2)$$

Here $|\vec{r}\rangle$ is the eigen-vector of the position operator $\vec{\mathcal{R}}$ and $|\psi\rangle$ is the vector of the superconducting state. This scalar product is by nature a complex number. Therefore, it fulfills the property of describing superconductivity from the outset. Furthermore, due to the wave function property, the quantity $|\psi(\vec{r})|^2$ corresponds to the *density* of superconducting particles.

The order parameter is not an independent variable as $\vec{B}$ and T, because it emerges as an additional parameter to the subgroup R_1. While $\vec{B}$ and T are fixed by external conditions, the field ψ is solely determined by the condition that the free energy takes its minimum value in the thermodynamic equilibrium. Therefore, the order parameter generally depends on $\vec{B}$ and T. It describes the origin of the phase transition, and it can be identified with a specific microscopic process.

The basic idea of the entire following considerations is that (as in any other Landau theory) the order parameter near the phase transition takes small values. This is used to expand the free energy around the critical point. The expansion coefficients depend on the temperature and magnetic induction. According to the principle of minimizing the thermodynamic potential in equilibrium, the order parameter is set in such a way that the free energy becomes minimal. From this minimization condition one can derive equations which determine the superconducting order parameter.

This property restricts the validity of the theory to temperatures relatively close to the transition point. On the other hand, it provides a rather complete and uniform description of phenomena related to the superconducting transition. The expansion parameters must be determined experimentally or calculated using appropriate microscopic many-particle theories. Furthermore, one should notice that the Ginzburg-Landau theory is a mean-field theory, i. e. the order parameter is considered as an average value. Fluctuations of the order parameter around its equilibrium value are not taken into account.

We start in section 4.1 with an idealized model of a superconducting material without external fields and surfaces. Measurable thermodynamic quantities are calculated and discussed. In section 4.2 the theory is generalized to an inhomogeneous system without external fields. The Ginzburg-Landau equations are solved for a simple one-dimensional model with an interface between a superconducting and normal conducting region. In this context, the Ginzburg-Landau coherence length is introduced.

Finally, in section 4.3 the Ginzburg-Landau equations are derived for a general superconducting system in the presence of external magnetic fields. The concepts of supercurrent and superfluid velocity are introduced. Moreover, the characteristic length scales entering the theory are discussed.

4.1 – Homogeneous superconductors

Let us start with the simplest possible realization of a superconductor: A homogeneous material without a magnetic field, which passes into the superconducting state below a critical temperature. In this special case we can set the magnetic induction equal to zero, $\vec{B} = 0$, and all material quantities have constant values in space. Thus, the free energy in the normal state is only a function of T, i. e. $F = F(T)$. Furthermore, for the homogeneous material it makes sense to introduce a density $f(T)$ of the free energy as follows,

$$F(T) = \int_{\mathcal{V}} \mathrm{d}^3 r' f(T) = V f(T).$$

In the following, we will use the density $f(T)$ instead of $F(T)$.

4.1.1 – Landau potential

Let us now turn to the superconducting state. We assume that the complex form (4.1) of the order parameter fully describes the property of superconductivity. This will be shown later by considering the electrical current density. Now, however, we first want to determine under which conditions such a state is thermodynamically stable.

A homogeneous superconductor is defined by the property of a spatially constant order parameter $\psi = |\psi|e^{i\Theta}$. In particular, in such a material, the superconducting particle density $|\psi|^2$ is constant. The free energy density in the superconducting state is a function of the temperature T and the constant order parameter ψ and we can write $f = f(T,\psi)$.

We start with the formulation of the explicit expression of the free energy density for the homogeneous material. As described in the previous section, we take advantage of the fact that near the phase transition, the order parameter takes on small values. Thus, close to the phase transition, a Taylor expansion of the free energy density with respect to $|\psi|$ is allowed. It reads

$$f(T,\psi) = f_n(T) + r(T)|\psi|^2 + \frac{u(T)}{2}|\psi|^4, \qquad (4.3)$$

where the expansion coefficients f_n, r, and u are functions of temperature. The coefficient $f_n(T)$ corresponds to the free energy density in the normal state. The function (4.3) in terms of ψ is called Landau potential. Terms of first and third order in ψ are not taken into account since they lead to unstable states at the transition point and at high temperatures.

The phenomenological parameters $r(T)$ and $u(T)$ are related to the specific material and have to be determined experimentally or by a microscopic theory. Nevertheless, if we take the parameters as given, we can gain a lot of insight into the temperature behavior of the order parameter close to the transition point. This will be explained in more detail further below. Finally, one should notice that the free energy density (4.3) is invariant under a continuous gauge transformation $\psi \rightarrow e^{i\alpha}\psi$ of the complex order parameter,

$$f(e^{i\alpha}\psi) = f(\psi).$$

It means that the free energy is symmetric with respect to the phase of the order parameter.

4.1.2 – Order parameter

The constant order parameter ψ is an auxiliary variable that serves to distinguish the superconducting state from the normal state. The square of its absolute value, $|\psi|^2 := n_s^*$, on the other hand, has an important physical significance: It describes the (spatially independent) density of superconducting particles and it corresponds to the density n_s of the superconducting 'liquid' introduced within the London theory. As will be shown later, in the Ginzburg-Landau theory n_s^* describes the density of *Cooper pairs*, whereas in the London theory n_s refers to the density of 'superconducting electrons'. Note that, in contrast to the London theory which is formulated to describe the low temperature limit $T \ll T_c$, the Ginzburg-Landau theory is valid for temperatures close to T_c.

Now, we search for the particular value of the superconducting order parameter which corresponds to the thermodynamic equilibrium. This is the value for which the free energy density $f(T,\psi)$ at a given temperature T (representing the thermodynamic equilibrium) becomes minimal. It is sufficient to minimize the free energy density with respect to $|\psi|^2$. The first and second derivative of (4.3) gives,

$$\frac{\partial f}{\partial(|\psi|^2)} = r(T) + u(T)|\psi|^2, \tag{4.4}$$

$$\frac{\partial^2 f}{\partial(|\psi|^2)^2} = u(T). \tag{4.5}$$

As is seen from the second derivative (4.5), f has a minimum only in the case $u(T) > 0$. Thus, in a superconducting material the parameter $u(T)$ must be always positive. Otherwise the system would be thermodynamically unstable. Setting the first derivative to zero, we obtain from (4.4) the equilibrium value of $|\psi|^2$ for which $f(T,\psi)$ becomes minimal,

$$|\psi|^2 = -\frac{r(T)}{u(T)} \quad \text{with} \quad u(T) > 0. \tag{4.6}$$

Note that r is changed by a variation of temperature. Therefore, depending on the value of r, we have to distinguish between two important cases.

Case $r > 0$

There is no finite solution for $|\psi|^2$ since the right hand side of (4.6) is negative. According to (4.3), for $r > 0$ and $u > 0$ the free energy density f becomes larger than the normal state value f_n for any finite value of $|\psi|^2$. Thus, we find

$$|\psi| = 0 \quad \text{for} \quad r(T) > 0, \tag{4.7}$$

and the system stays in the normal state as long as r has a positive value.

Case $r < 0$

In this case the free energy is minimal if the order parameter takes the finite value

$$|\psi| = \sqrt{\frac{|r(T)|}{u(T)}} \quad \text{for} \quad r(T) < 0. \tag{4.8}$$

Consequently, the system prefers the superconducting state rather than the normal state. The new state is described by the finite order parameter ψ which depends on temperature T via the phenomenological parameters $r(T)$ and $u(T)$. It has a lower free energy than the normal state. The critical temperature T_c is determined by $r(T_c) = 0$.

Next, we elaborate the temperature dependence of the order parameter near the transition point. This regime is accessible through a Taylor expansion with respect to T around T_c,

$$r(T) = \tilde{r}_0(T - T_c) + \ldots, \tag{4.9}$$

$$u(T) = \tilde{u}_0 + \tilde{u}_1(T - T_c) + \ldots. \tag{4.10}$$

The coefficients $\tilde{r}_0$, $\tilde{u}_0$, and $\tilde{u}_1$ are temperature-independent. The expansions (4.9) and (4.10) allow us to calculate the temperature dependence: At first, equations (4.7) and (4.9) lead to $|\psi| = 0$ for $T > T_c$. In the case of temperatures below the critical point, $T < T_c$, we have to consider equation (4.8). Taking in the Taylor expansions (4.9) and (4.10) only the lowest order in $(T - T_c)$ we obtain the temperature behavior

$$|\psi| \propto \sqrt{T_c - T}, \tag{4.11}$$

which is then valid for temperatures $T < T_c$ close to the transition point. Thus, the result (4.11) describes the temperature behavior of the order parameter near the phase transition (as is schematically shown by the solid line in Figure 4.1). The power law behavior $\propto (T_c - T)^\beta$ with an exponent $\beta = 1/2$ is typical of a mean field theory considered here.

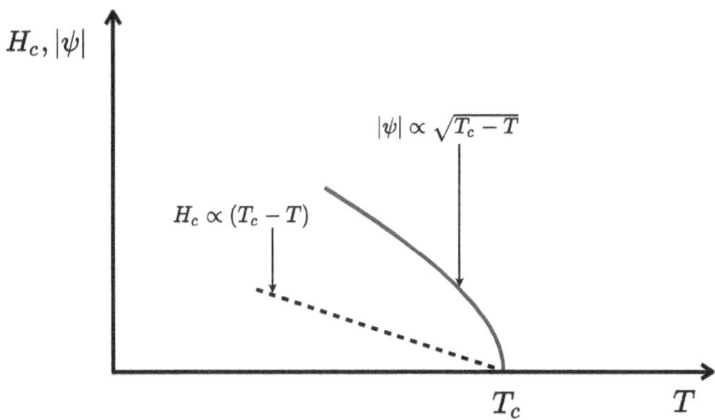

Figure 4.1: Temperature behavior of the superconducting order parameter $|\psi|$ and the critical magnetic field H_c close to the transition temperature T_c.

In summary, the temperature dependence of the order parameter is as follows. It is zero in the normal phase, $\psi = 0$ for $T > T_c$, and it becomes non-zero in the superconducting phase, with the temperature dependence $|\psi| \propto \sqrt{T_c - T}$ for $T < T_c$. Close to the phase transition the order parameter takes small values.

Finally, one should mention that r and u are quantities which are relatively difficult to measure directly by experiments. However, it is easy to express these parameters in terms of observables which are more convenient to measure. These are the critical field H_c and the density of superconducting particles $n_s^* = |\psi|^2 = |r|/u$. Both quantities show a linear dependence $\propto (T_c - T)$ near the critical point (dashed line in Figure 4.1). For $|\psi|^2$, this can be easily recognized from equation (4.11). The linear temperature behavior of H_c is justified later.

4.1.3 – Specific heat capacity

Following the above ideas, we are now able to investigate the temperature dependence of the specific heat capacity near the phase transition. The specific heat capacity is experimentally very well accessible and shows a characteristic jump when the material undergoes the transition from the superconducting to the normal state (compare Figure 1.7). With the help of our theory developed so far, we can explain this jump as follows.

We calculate the specific heat close to the transition point using the *entropy difference* between the superconducting and the normal state. According to (3.13), the entropy of a thermodynamic system is given by the negative partial derivative of the free energy with respect to temperature. Thus, the difference $\Delta s = s_s - s_n$ between the entropy values per volume in the superconducting state s_s and the normal state s_n is given by

$$\Delta s = s_s - s_n = -\frac{\partial \Delta f(T)}{\partial T},$$

where the temperature dependence of the free energy difference per volume, $\Delta f(T) = f(T) - f_n(T)$, is determined by the parameters $r(T)$ and $u(T)$ in the Landau potential (4.3). In addition, of course, $|\psi|^2$ is still dependent on T, but we can also replace this quantity with r and u, with the help of (4.8). We obtain

$$\Delta f = f - f_n = r\,|\psi|^2 + \frac{u}{2}|\psi|^4 = r\,\frac{|r|}{u} + \frac{u}{2}\frac{|r|^2}{u^2} = -\frac{|r|^2}{2u}.$$
$$(4.12)$$

We have used that in the superconducting state r is negative, i. e. we can write $r = -|r|$.

We are again interested in the behavior of the material within a *narrow temperature range* close to the transition temperature. Therefore, we can use the expansions (4.9) and (4.10) to find the temperature dependence $\Delta f(T)$. Using (4.12), considering again the lowest order in $(T - T_c)$, and forming the temperature derivative, we obtain

$$\Delta s = \frac{\partial}{\partial T}\frac{|r|^2}{2u} \approx \frac{\tilde{r}_0^2}{2\tilde{u}_0}\frac{\partial}{\partial T}(T - T_c)^2 = -\frac{\tilde{r}_0^2}{\tilde{u}_0}(T_c - T). \quad (4.13)$$

From this equation we may draw an important conclusion. In the superconducting state, where $T < T_c$, the entropy difference is negative, i. e. the entropy in the superconducting state is lower than in the normal state. From the concepts of thermodynamics we know that the entropy represents the 'degree of disorder' of a thermodynamic system. Consequently, the superconducting state has a higher 'degree of order' than the normal state. This property is represented by the emergence of the superconducting order parameter ψ.

The specific heat per volume is defined as the infinitesimal amount of heat (per volume) supplied to the system, divided by the associated temperature change, $c = \delta q/dT$. The amount of heat is associated with the change in entropy via the second law of thermodynamics, $\delta q = Tds$. Using this equation, the specific heat per volume is obtained from the temperature derivative of the entropy per volume, $c = T(\partial s/\partial T)$. Thus, the difference $\Delta c = c_s - c_n$ between the specific heat values c_s and c_n in the superconducting and normal state can be calculated as follows,

$$\Delta c = c_s - c_n = T\frac{\partial \Delta s}{\partial T}.$$

For temperatures close to the transition, Δs is given by (4.13), and the derivative can be calculated explicitly. The result is a linear T-dependence,

$$\Delta c \approx \frac{\tilde{r}_0^2}{\tilde{u}_0} T, \qquad (4.14)$$

in the vicinity of the transition point in the superconducting state. This result explains the discontinuity in the electronic part of the specific heat, which was found by many experiments (compare section 1.4). Assuming that the superconducting state is a phenomenon caused by the conduction electrons we can attribute the value Δc to the electronic part of the specific heat. The normal state heat capacity c_n follows a linear T-behavior, $c_n = \gamma T$. Consequently, according to (4.14), the specific heat c_s in the *superconducting state* follows a linear behavior as well,

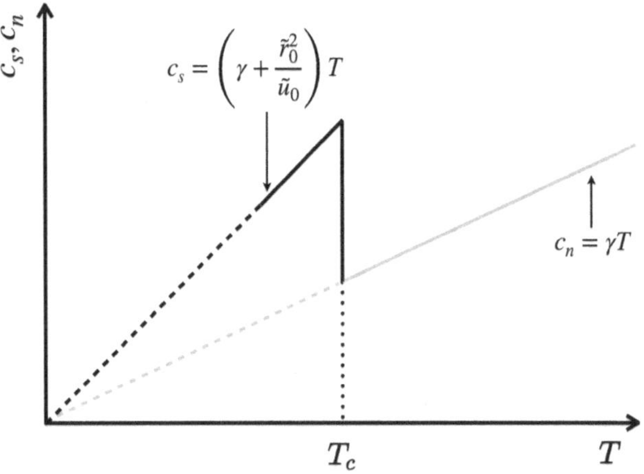

Figure 4.2: Linear temperature behavior of the specific heat in the superconducting state (black line) close to the transition temperature T_c. The normal state specific heat (gray line) is also linear but has a smaller slope. Since both c_s and c_n approach to zero for $T \to 0$ (dashed lines), a discontinuity must appear at T_c.

$$c_s \approx c_n + \frac{\tilde{r}_0^2}{\tilde{u}_0}T = \left(\gamma + \frac{\tilde{r}_0^2}{\tilde{u}_0}\right)T,$$

but with a different slope. The T-dependence of c_s and c_n is illustrated in Figure 4.2. Both $c_s(T)$ and $c_n(T)$ are linear functions with different slopes. The slope of c_s is by the value of $\tilde{r}_0^2/\tilde{u}_0$ larger than the slope of c_n. Both lines end in the origin, i. e.,

$$c_s(T \to 0) = c_n(T \to 0) = 0.$$

Such a behavior implies that the specific heat has a discontinuity at the critical temperature T_c, in agreement with the experimental finding schematically shown by Figure 1.7. The height of the jump is approximately

$$\Delta c(T = T_c) \approx (\tilde{r}_0^2 T_c)/\tilde{u}_0.$$

4.2 – Inhomogeneous superconductors

We now generalize our theoretical description of a superconducting material to a system in which the superconducting order parameter may change in space. The perhaps simplest realization of such a system might be a superconducting material that is bounded to a normal conducting material. Of particular interest in this context is the spatial variation of the superconducting order parameter nearby the boundary. We will show that the order parameter varies within a typical length scale which is determined by the coherence length ξ.

At first, let us generalize the expression (4.3) for the free energy to a situation where the order parameter ψ may change in space, $\psi \to \psi(\vec{r})$, i. e. it becomes a function of the position vector $\vec{r}$. The postulate of the Ginzburg-Landau theory is that the superconducting order parameter corresponds to the quantum mechanical wave function (4.2) describing the state of the superconducting condensate. As any other spatially dependent wave function, the free energy must contain a term proportional to

$$\int_{\mathcal{V}} d^3r \, \frac{\hbar^2}{2m^*} |\vec{\nabla}\psi(\vec{r})|^2 := \int_{\mathcal{V}} d^3r \, \frac{\hbar^2}{2m^*} \vec{\nabla}[\psi(\vec{r})] \cdot \vec{\nabla}[\psi^*(\vec{r})],$$

$$(4.15)$$

which describes the kinetic energy of the many-particle wave function. The integral runs over the volume $\mathcal{V}$ of the superconductor. Note that on the right hand side of the equation the Nabla operator acts solely on the particular function inside the brackets. The effective mass of the superconducting particles is denoted by m^*. Note that in quantum mechanics the operator $(\hbar/i)\vec{\nabla}$ has the meaning of the one-particle momentum operator in spatial representation.

The extension of (4.3) to a spatially dependent order parameter is simply the replacement of the volume with a spatial integration over the second and third term of (4.3) and the introduction of a kinetic energy (4.15). Thus, assigning the superconducting particle a mass m^*, the quantum mechanical formulation of the total free energy for the superconducting condensate reads

$$F[T,\psi] = F_n + \int_{\mathcal{V}} d^3r \left(\frac{\hbar^2}{2m^*} |\vec{\nabla}\psi(\vec{r})|^2 \right.$$
$$\left. + r|\psi(\vec{r})|^2 + \frac{u}{2}|\psi(\vec{r})|^4 \right), \qquad (4.16)$$

where F_n is the free energy of the normal state. The phenomenological parameters $u = u(T)$ and $r = r(T)$ depend on temperature and may, in principle, also vary in space which is not considered here for the sake of simplicity. Furthermore, also the free energy in the normal state $F_n = F_n(T)$ generally depends on temperature.

As already introduced above, the notation $F[T,\psi]$ means that the free energy F is a *functional* with respect to the complex field $\psi(\vec{r})$, i. e. a number $F[T,\psi]$ is assigned to the field $\psi(\vec{r})$ and the temperature T. As can be seen in (4.16), this functional is invariant under a *global* gauge transformation,

$$\psi(\vec{r}) \to e^{i\alpha}\psi(\vec{r}),$$

where α is constant. Since the order parameter generally changes spatially, but the gauge transformation can only be carried out globally, one must usually take into account both the order parameter ψ and its conjugate ψ^*.

4.2.1 – Minimization of the free energy

We search for the particular order parameter function $\psi(\vec{r})$ for which the free energy (4.16) becomes minimal. This special solution is taken by the system in thermodynamic equilibrium and it is the central aim of the Ginzburg-Landau theory to find this solution with respect to the given boundary conditions (surface of the superconductor).

At first, we introduce an infinitesimal variation of the wave function $\psi(\vec{r})$ by considering the slightly changed function

$$\psi(\vec{r}) + \delta\psi(\vec{r}), \tag{4.17}$$

where $\delta\psi(\vec{r})$ is an arbitrary function taking infinitesimally small values at any $\vec{r}$, except on the surface of the superconductor where $\delta\psi(\vec{r})$ is set equal to zero. The function (4.17) is intended as an infinitesimally small variation of $\psi(\vec{r})$, which still meets the boundary conditions.

The basic idea of the minimization method is as follows. We use the variation function (4.17) to calculate the free energy from equation (4.16) and compare the obtained result with the corresponding value of the free energy $F[T, \psi]$ without variation. Technically, we calculate the difference

$$\delta F = F[T, \psi + \delta\psi] - F[T, \psi], \tag{4.18}$$

where $\psi + \delta\psi$ denotes the function (4.17), and determine the particular function $\psi(\vec{r})$ for which $\delta F = 0$ for an arbitrary chosen variation function $\delta\psi(\vec{r})$.

At first, let us evaluate $F[T, \psi + \delta\psi]$. Inserting (4.17) in (4.18), we find

$$F[T, \psi + \delta\psi] = F_n + \int_{\mathcal{V}} d^3r \left[\frac{\hbar^2}{2m^*} \vec{\nabla}(\psi + \delta\psi) \cdot \vec{\nabla}(\psi^* + \delta\psi^*) \right.$$
$$\left. + r(\psi + \delta\psi)(\psi^* + \delta\psi^*) + \frac{u}{2}(\psi + \delta\psi)^2(\psi^* + \delta\psi^*)^2 \right].$$

Since the variation $\delta\psi$ is infinitesimally small, we are allowed to neglect all terms of second and higher order in the variation. Thus, the remaining terms are $F[T, \psi]$ ('zero' order) and terms proportional to $\delta\psi$ and $\delta\psi^*$. We find

$$F[T, \psi + \delta\psi] = F[T, \psi] + \int_{\mathcal{V}} d^3r \left[\frac{\hbar^2}{2m^*} \vec{\nabla}(\psi) \cdot \vec{\nabla}(\delta\psi^*) + \right.$$
$$\frac{\hbar^2}{2m^*} \vec{\nabla}(\delta\psi) \cdot \vec{\nabla}(\psi^*) + r(\psi\delta\psi^* + \psi^*\delta\psi) +$$
$$\left. u|\psi|^2(\psi\delta\psi^* + \psi^*\delta\psi) \right]. \tag{4.19}$$

Next, we reformulate the first and second volume integral such that the Nabla operator only applies to the fields ψ and ψ^* and not to the variations. We use the general product rule of arbitrary scalar fields $\alpha(\vec{r})$ and arbitrary vector fields $\vec{A}(\vec{r})$,

$$\vec{\nabla} \cdot (\alpha\vec{A}) = \vec{A} \cdot \vec{\nabla}\alpha + \alpha\vec{\nabla} \cdot \vec{A}, \tag{4.20}$$

to rewrite the first volume integral in (4.19) as follows

$$\int_{\mathcal{V}} d^3r \vec{\nabla}(\psi) \cdot \vec{\nabla}(\delta\psi^*) = \int_{\mathcal{V}} d^3r \left[\vec{\nabla} \cdot (\delta\psi^* \vec{\nabla}\psi) - \right.$$
$$\left. \delta\psi^* \vec{\nabla} \cdot \vec{\nabla}\psi \right]. \tag{4.21}$$

The first term is an integral over the divergence of a vector field. According to the Gauss theorem, it can be replaced with an integral over the surface $\mathcal{S}$ of the superconducting material,

$$\int_{\mathscr{V}} d^3r\,\vec{\nabla}\cdot(\delta\psi^*\,\vec{\nabla}\psi) = \int_{\mathscr{S}} d\vec{f}\cdot(\delta\psi^*\,\vec{\nabla}\psi) = 0,$$

where $d\vec{f}$ is the surface element at a position $\vec{r}$ on the surface $\mathscr{S}$. The surface integral disappears because we have set the variation $\delta\psi^*$ equal to zero on the entire surface. Thus, the formula (4.21) of the volume integral simplifies to

$$\int_{\mathscr{V}} d^3r\,\vec{\nabla}(\psi)\cdot\vec{\nabla}(\delta\psi^*) = -\int_{\mathscr{V}} d^3r\,\delta\psi^*\,\vec{\nabla}^2\psi. \tag{4.22}$$

It is seen that the Laplace operator

$$\vec{\nabla}^2 = \vec{\nabla}\cdot\vec{\nabla} = \frac{\partial^2}{\partial x^2} + \frac{\partial^2}{\partial y^2} + \frac{\partial^2}{\partial z^2}$$

acts on the field ψ while the variation $\delta\psi^*$ now appears as a pre-factor.

The second volume integral in (4.19) is the conjugate of the one just discussed and can be treated in the same way. Thus, inserting (4.22) in (4.19), we obtain the following expression for the variation δF of the free energy,

$$\delta F = \int_{\mathscr{V}} d^3r\,\delta\psi^*\left(-\frac{\hbar^2}{2m^*}\vec{\nabla}^2\psi + r\psi + u|\psi|^2\psi\right)$$
$$+ \int_{\mathscr{V}} d^3r\,\delta\psi\left(-\frac{\hbar^2}{2m^*}\vec{\nabla}^2\psi^* + r\psi^* + u|\psi|^2\psi^*\right). \tag{4.23}$$

The free energy F is minimal if $\delta F = 0$ with respect to an arbitrary variation $\delta\psi$ (and $\delta\psi^*$). Consequently, the particular solutions $\psi(\vec{r})$ and $\psi^*(\vec{r})$ for which $F[T,\psi]$ becomes minimal are determined by the property that both brackets in the volume integrals in (4.23) are zero. Thus, the order parameter functions in thermodynamic equilibrium fulfill the differential equations

$$-\frac{\hbar^2}{2m^*}\vec{\nabla}^2\psi + r\psi + u|\psi|^2\psi = 0, \tag{4.24}$$

$$-\frac{\hbar^2}{2m^*}\,\vec{\nabla}^2\psi^* + r\psi^* + u\,|\psi|^2\psi^* = 0. \tag{4.25}$$

These two equations are the time-independent Schrödinger equations for the superconducting order parameter $\psi(\vec{r})$ and its complex conjugate $\psi^*(\vec{r})$. The second equation is the complex conjugate of the first equation. The two equations are equivalent since the parameters u and r are real numbers. Differences between ψ and ψ^* may enter only through the boundary conditions. Thus, if the order parameter is real on the entire boundary, it can be treated as a real scalar field, $\psi(\vec{r}) = \psi^*(\vec{r})$, in the entire material. In this case the equation (4.24) is the basic differential equation which determines the superconducting order parameter in a field-free material. It is called the Ginzburg-Landau equation.

4.2.2 – Functional derivative

Before presenting a solution of the Ginzburg-Landau equation for a simple case of an inhomogeneous superconductor, let us quickly discuss an alternative derivation of (4.24), based on the functional derivative of the free energy. The functional derivative was defined above by the equations (3.8) and (3.6).

The advantage of the concept of functional derivatives is that the minimization of the free energy can be done relatively quickly by calculating the first derivatives of the free energy density with respect to ψ and ψ^*. We show that within such a procedure the Ginzburg-Landau equation is obtained.

At first, we write the free energy (4.16) formally like (3.6) in terms of a volume integral of the energy density which is a function of ψ and ψ^*,

$$F[T,\psi] = F_n + \int_{\mathcal{V}} d^3r\, f(T,\psi(\vec{r}),\psi^*(\vec{r})). \tag{4.26}$$

The minimization of F is done by forming the functional derivatives of F with respect to ψ and ψ^* and setting them to zero. Genera-

lizing (3.8) to a situation with two variables, we obtain the following equation for the minimization,

$$\frac{\delta F}{\delta \psi(\vec{r})} := \frac{\partial f(T, \psi, \psi^*)}{\partial \psi} = 0. \tag{4.27}$$

Before evaluating the partial derivative of f with respect to ψ,
the integral should be written such that ψ appears explicitly and
that no differential operator acts on ψ. We show this starting from
the Ginzburg-Landau potential (4.16),

$$F[T, \psi] = F_n + \int_{\mathscr{V}} d^3r \left[\frac{\hbar^2}{2m^*} \vec{\nabla}[\psi(\vec{r})] \cdot \vec{\nabla}[\psi^*(\vec{r})] \right. $$
$$\left. + r|\psi(\vec{r})|^2 + \frac{u}{2}|\psi(\vec{r})|^4 \right],$$

after (4.15) was used. Again, we reformulate the first volume integral such that the first Nabla operator acting on ψ is 'shifted' to act
on ψ^*. The procedure is the same as finding relation (4.22) by applying rules of vector analysis and choosing a surface outside the
superconductor where $\psi = 0$. After reformulating the first volume
integral in that way we find

$$F[T, \psi] = F_n + \int_{\mathscr{V}} d^3r \left[-\frac{\hbar^2}{2m^*} \psi(\vec{r}) \vec{\nabla}^2 \psi^*(\vec{r}) \right. $$
$$\left. + r|\psi(\vec{r})|^2 + \frac{u}{2}|\psi(\vec{r})|^4 \right],$$

It is an expression of the form of (4.26) with the density function

$$f(T, \psi, \psi^*) = -\frac{\hbar^2}{2m^*} \psi \vec{\nabla}^2 \psi^* + r(T) \psi \psi^* + \frac{u(T)}{2} \psi^2 (\psi^*)^2. \tag{4.28}$$

As required, no differential operator acts on the field ψ. Now we
may calculate the functional derivative (4.27) by simply evaluating
the partial derivative of (4.28) with respect to ψ. Setting the func-

tional derivative to zero (minimization of the free energy), we immediately obtain the non-linear Schrödinger equation (4.25) for ψ^*,

$$\frac{\partial f(T,\psi,\psi^*)}{\partial \psi} = -\frac{\hbar^2}{2m^*}\vec{\nabla}^2\psi^* + r(T)\psi^* + u(T)\psi(\psi^*)^2 = 0.$$

$$(4.29)$$

The equation for ψ is found from the functional derivative with respect to ψ^* which leads us to the complex conjugate of equation (4.29).

4.2.3 – Ginzburg-Landau coherence length

Now we solve the Ginzburg-Landau equation (4.24) for a simple case of a superconducting material which is bounded to a normal conducting material. As will be shown in the following, the solution for ψ shows a characteristic behavior nearby the boundary and changes in space within a certain length scale, the Ginzburg-Landau coherence length.

We consider an infinite flat interface lying in the y-z plane and separating a normal metal in the region $x < 0$ from a superconductor in the region $x > 0$. The symmetry with respect to y and z reduces equation (4.24) to a one-dimensional problem. The order parameter ψ depends only on x and the Laplace operator $\vec{\nabla}^2$ can be replaced with the second derivative with respect to x,

$$-\frac{\hbar^2}{2m^*}\frac{d^2\psi}{dx^2} + r\psi + u|\psi|^2\psi = 0. \qquad (4.30)$$

We choose a gauge in which ψ is real. Furthermore, it is convenient to refer ψ to its bulk value, $\psi_0 = \sqrt{-r/u}$, as calculated in section 4.1. We write $\psi(x)$ in the form $\psi(x) = \psi_0 f(x)$, where $f(x)$ is a dimensionless function with the property $f(x \to \infty) = 1$. On the boundary $x = 0$ to the normal state material the order parameter is zero,

$$\psi(x = 0) = f(x = 0) = 0.$$

Inserting in (4.30) $\psi(x) = \sqrt{-r/u}f(x)$, we obtain at first

$$-\frac{\hbar^2}{2m^*}\sqrt{-\frac{r}{u}}\frac{d^2 f(x)}{dx^2} + r\sqrt{-\frac{r}{u}}f(x) + u\left(-\frac{r}{u}\right)^{\frac{3}{2}} f^3(x) = 0.$$

Division by $(r\sqrt{-r/u})$ leads us to

$$-\frac{\hbar^2}{2m^*r}\frac{d^2 f(x)}{dx^2} + f(x) + \frac{u}{r}\left(-\frac{r}{u}\right) f^3(x) = 0.$$

With the temperature-dependent coherence length

$$\xi(T) := \sqrt{\frac{\hbar^2}{2m^*|r(T)|}}, \tag{4.31}$$

we find a more compact form of the above differential equation,

$$-\xi^2(T)\frac{d^2 f(x)}{dx^2} - f(x) + f^3(x) = 0. \tag{4.32}$$

From the first term one can see that the Ginzburg-Landau cohe-rence length $\xi(T)$ appears as a characteristic length scale in which the order parameter ψ changes.

The temperature dependence $\xi(T)$ of the Ginzburg-Landau co-herence length close to the transition temperature T_c can be found from $|r(T)| \propto (T_c - T)$ (compare (4.9)). We obtain

$$\xi(T) \propto \frac{1}{\sqrt{T_c - T}} \tag{4.33}$$

close to T_c. Thus, the mean-field critical exponent is (-1/2). Result (4.33) has an important consequence: The coherence length diver-ges at the critical temperature! This behavior can be interpreted as follows. As will be shown below, the coherence length describes the characteristic length scale in which the order parameter takes smal-ler values than its bulk value ψ_0. This reduction appears within a certain distance (of the order of ξ) to the normal conducting boundary. Thus, very close to the transition to the normal state, this length scale is spread over the whole volume of the superconductor.

Let us now solve the differential equation (4.32) assuming that in the normal metal the order parameter vanishes identically. In addition, the order parameter should be continuous at the interface. At first, we multiply (4.32) by the first derivative f' of f,

$$-\xi^2 f'' f' - f f' + f^3 f' = 0,$$

and we apply the product rule,

$$-\xi^2 \frac{1}{2} \frac{d}{dx} f'^2(x) - \frac{1}{2} \frac{d}{dx} f^2(x) + \frac{1}{4} \frac{d}{dx} f^4(x) = 0.$$

We integrate this equation from some value x to infinity and find after multiplication by 2,

$$-\xi^2 \left[f'^2(\infty) - f'^2(x) \right] - \left[f^2(\infty) - f^2(x) \right] + \frac{1}{2} \left[f^4(\infty) - f^4(x) \right] = 0.$$

In the limit $x \to \infty$, the functions f and f' approach the values $f(\infty) = 1$ and $f'(\infty) = 0$, respectively. This leads us to the first-order differential equation

$$\xi^2 f'^2(x) + f^2(x) - \frac{1}{2} f^4(x) - \frac{1}{2} = 0,$$

or

$$\xi^2 f'^2(x) = \frac{1}{2} \left[1 - f^2(x) \right]^2.$$

Further simplification is possible by forming the square root,

$$f'(x) = \pm \frac{1}{\xi\sqrt{2}} \left[1 - f^2(x) \right].$$

We solve this differential equation by separation of variables and integration between the boundary $x = 0$, where $f(0) = 0$, and some value x,

$$\int_0^{f(x)} \frac{1}{1-y^2}\,dy = \pm\frac{1}{\xi\sqrt{2}} \int_0^x dx' = \pm\frac{x}{\xi\sqrt{2}}. \qquad (4.34)$$

Finally, using the standard integral

$$\int 1/(1-y^2)\,dy = \operatorname{artanh}(y)$$

with $\operatorname{artanh}(0) = 0$ and solving the resulting equation for $f(x)$, we obtain

$$f(x) = \pm\tanh\left(\frac{x}{\xi\sqrt{2}}\right).$$

Since only the density $n_s(x) = |\psi(x)|^2$ of superconducting particles is relevant for experiments both solutions of $f(x)$ are equivalent. For convenience, we choose the positive solution. With $\psi = \psi_0 f(x)$ we obtain the final result for the superconducting order parameter,

$$\boxed{\psi(x) = \psi_0 \tanh\left(\frac{x}{\xi\sqrt{2}}\right)} \qquad (4.35)$$

for $x > 0$, and $\psi(x) = 0$ for $x < 0$.

The result is plotted in Figure 4.3. In the normal conducting region, $x < 0$, the order parameter is zero. In the superconducting region, $x > 0$, its value increases within a characteristic length scale given by the Ginzburg-Landau coherence length ξ. Deep inside the superconductor, i. e. in the limit $x \to \infty$, $\psi(x)$ approaches its bulk value ψ_0. Consequently, the superconducting particle density n_s is reduced within a certain region near the surface. Note that the coherence length diverges at the transition point T_c, leading to a reduction of n_s throughout the whole volume of the superconductor. This marks the transition to the normal state.

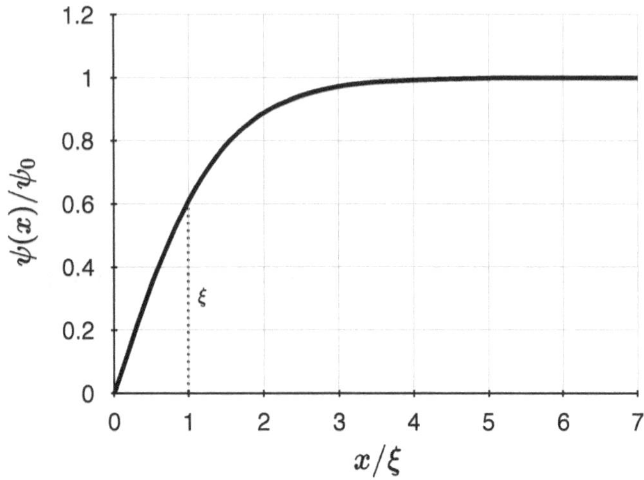

Figure 4.3: Superconducting order parameter according to the solution (4.35) as a function of the distance x from the boundary (placed at $x = 0$) to a normal conducting material ($x < 0$) characterized by $\psi = 0$. The order parameter changes significantly within a characteristic length scale determined by the Ginzburg-Landau coherence length ξ (dotted line). Deep inside the superconductor, ψ approaches its bulk value ψ_0.

4.2.4 – Proximity effect

Using the above consideration we now study the effect of a finite Cooper pair density in the normal conducting region close to a superconductor. The phenomenon describes a continuation of finite order parameter values to the normal metal region. In experiments this effect is known as the proximity effect.

At first, let us point out again that the particular order parameter value $\psi(x = 0)$ at the interface is determined by the boundary condition which must be fixed before the Ginzburg-Landau equation is solved. So far we have considered the boundary condition $\psi(x = 0) = 0$. This means, on the boundary the material is purely normal conducting. We now generalize our treatment by considering a different boundary condition, $\psi(0) = C$ with $0 < C < \psi_0$, which allows for a finite Cooper pair density at the interface between the superconductor and the normal metal. In the following, we will show that this somewhat more realistic assumption actually gives rise to the proximity effect.

Let us start with the behavior of ψ inside the superconductor. Thereby, the new boundary condition enters the integral (4.34). Here, the lower boundary of the integral is to be replaced with $f(0) = C/\psi_0$,

$$\int_{C/\psi_0}^{f(x)} \frac{1}{1-y^2} dy = \operatorname{artanh}(f(x)) - \operatorname{artanh}\left(\frac{C}{\psi_0}\right) = \pm \frac{x}{\xi\sqrt{2}}.$$

Solving this equation for $\psi(x) = \psi_0 f(x)$ leads to the result for the order parameter in the superconducting region $x > 0$,

$$\psi(x) = \psi_0 \tanh\left(\frac{x}{\xi\sqrt{2}} + \tilde{C}\right), \tag{4.36}$$

where $\tilde{C} = \operatorname{artanh}(C/\psi_0)$. The qualitative course of this function is comparable to the solution (4.35) with the previous boundary condition. The difference is only the finite value at $x = 0$.

Now we turn to the study of ψ in the *normal conducting material* in the region $x < 0$. In particular, we show that a finite Cooper pair density n_s occurs there despite the presence of the metallic state. At first, let us have a look on the non-linear Schrödinger equation

$$-\frac{\hbar^2}{2m^*} \vec{\nabla}^2 \psi + r\psi + u|\psi|^2 \psi = 0 \tag{4.37}$$

and see whether a non-zero solution for ψ is possible. A normal conducting material is defined by the property that the material is normal conducting in the bulk. This means the parameter r is positive in the region $x < 0$. We define $r = r_n > 0$ for $x < 0$. The second parameter u is always positive since according to (4.10) its lowest order is temperature independent. In the normal state, we set $u = u_n$ where $u_n \approx \tilde{u}_0 > 0$.

We will now show that in the normal state with $r_n > 0$ and $u_n > 0$ near the boundary Cooper pairs can still exist ($\psi \neq 0$). Introducing a positive function $f(x) > 0$ with

$$\psi(x) = f(x)\sqrt{\frac{r_n}{u_n}} \qquad (4.38)$$

and replacing in (4.37) $\psi(x)$ with $f(x)$, we find a differential equation for the function $f(x)$:

$$-\frac{\hbar^2}{2m^*}\sqrt{\frac{r_n}{u_n}}\frac{\mathrm{d}^2 f(x)}{\mathrm{d}x^2} + r_n\sqrt{\frac{r_n}{u_n}}f(x) + u_n\left(\frac{r_n}{u_n}\right)^{\frac{3}{2}} f^3(x) = 0.$$

Dividing this equation by $r_n\sqrt{r_n/u_n}$, we obtain

$$-\xi_n^2(T)\frac{\mathrm{d}^2 f(x)}{\mathrm{d}x^2} + f(x) + f^3(x) = 0,$$

where

$$\xi_n(T) = \frac{\hbar}{\sqrt{2m^* r_n(T)}}.$$

Assuming that the density of superconducting particles in the normal material is relatively small, i. e. $f(x) \ll 1$, we neglect the third order term,

$$-\xi_n^2(T)\frac{\mathrm{d}^2 f(x)}{\mathrm{d}x^2} + f(x) \approx 0,$$

for $f(x) \ll 1$. Multiplication by f' and reformulation of the derivatives gives

$$-\xi_n^2\frac{1}{2}\frac{\mathrm{d}}{\mathrm{d}x}[f'(x)]^2 + \frac{1}{2}\frac{\mathrm{d}}{\mathrm{d}x}f^2(x) = 0.$$

Integrating this equation from $(-\infty)$ to x and taking into account the limiting values $f(-\infty) = f'(-\infty) = 0$, we obtain a first order differential equation

$$-\xi_n^2[f'(x)]^2 + f^2(x) = 0,$$

or $f(x) = \pm \xi_n f'(x)$. Since $f(x) \geq 0$ and $\xi_n > 0$, the differential equation with the positive sign, $f(x) = \xi_n f'(x)$, is physically relevant. The equation with the negative sign, however, would give rise to a steady increase of $f(x)$ in the metallic region and in the limit $x \to -\infty$ the corresponding solution would be superconducting.

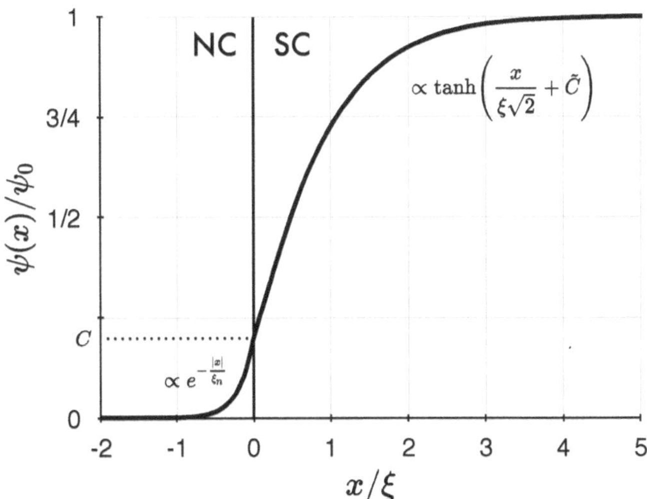

Figure 4.4: Superconducting order parameter as a function of x according to the solutions (4.36) and (4.39). The boundary between normal conducting material (NC) and superconducting material (SC) is placed at $x = 0$. In contrast to Figure 4.3, the boundary condition considered here, $\psi(0) = C$, gives rise to a finite Cooper pair density on the interface. The parameters chosen in the figure are $\xi_n/\xi = \tilde{C}/\xi = 0.2$.

Thus, we continue with the differential equation $f(x) = \xi_n f'(x)$ and solve it using separation of variables. Integration between a position $x < 0$ in the metallic region and the surface leads us to

$$\int_x^0 dx' = \xi_n \int_{f(x)}^{f(0)} \frac{dy}{y} = \xi_n \ln \left| \frac{f(0)}{f(x)} \right|.$$

On the surface we have $\psi(0) = C$ or $f(0) = C\sqrt{u/r_n}$, and therefore we get

$$-x = \xi_n \ln\left(\frac{C}{f(x)}\sqrt{\frac{u_n}{r_n}}\right) = |x|$$

or

$$\frac{C}{f(x)}\sqrt{\frac{u_n}{r_n}} = e^{\frac{|x|}{\xi_n}}.$$

Replacing again $f(x)$ with $\psi(x)$ by use of (4.38), we obtain the final result for the order parameter function in the normal conducting region $x < 0$,

$$\psi(x) = C e^{-\frac{|x|}{\xi_n}}. \tag{4.39}$$

This result explains the proximity effect. The behavior of the order parameter as a function of x is shown in Figure 4.4. If a normal conducting material (in the bulk) is connected to a superconductor, a finite density $|\psi|^2$ of superconducting particles (Cooper pairs) is measured in a narrow region close to the interface between both materials. Inside this region, the density decays exponentially within the characteristic length scale ξ_n. Note that, in general, ξ_n differs from the Ginzburg-Landau coherence length since it arises from phenomenological parameters u_n and r_n of the normal metal.

For very clean metals the proximity effect appears in the order of several hundredths of micrometers. It is particularly used in Josephson junctions where Cooper pairs tunnel through normal conducting barriers.

4.2.5 – Loss of energy due to the boundary

The reduction of the order parameter compared to its bulk value, which occurs near the boundary of the superconductor, should lead to a higher total energy. This effect is important in the context of the formation of domain walls. In the following, we will calculate this energy increase for the system with translational symmetry and show that it is directly related to the coherence length and the critical magnetic field.

The total free energy can be calculated from (4.16) by taking into account that $\psi(x)$ solely depends on x. We find

$$F = F_n + A \int_0^\infty dx \left(-\frac{\hbar^2}{2m^*} \psi \frac{d^2}{dx^2} \psi^* + r|\psi|^2 + \frac{u}{2}|\psi(\vec{r})|^4 \right),$$

(4.40)

where A is the area of the interface between normal metal and superconducting material. The order parameter function $\psi(x)$ is given by solution (4.35) and fulfills the Ginzburg-Landau equation (4.30).

Next, we multiply the Ginzburg-Landau equation by ψ^*,

$$-\frac{\hbar^2}{2m^*} \psi^* \frac{d^2 \psi}{dx^2} + r|\psi|^2 + u|\psi|^2|\psi|^2 = 0,$$

and we use this equation to eliminate the second derivative in (4.40). We obtain

$$F = F_n - \frac{Au}{2} \int_0^\infty dx\, |\psi|^4.$$

(4.41)

The integral in (4.41) describes the energy gain of the superconductor related to the energy F_n of the material in the normal state.

We are now interested in the increase of energy due to the reduction of the order parameter near the surface of the superconductor. It is the energy difference

$$\Delta F := F - F_0$$

(4.42)

between the free energy F of the material with the finite coherence length ξ, given by (4.41), and the free energy

$$F_0 = F_n - \frac{Au}{2} \int_0^\infty dx\, |\psi_0|^4$$

of a material with a constant ψ_0 up to the boundary (which corresponds to the homogeneous solution with $\xi = 0$). Thus, the energy difference per area can be calculated by the integral

$$\frac{\Delta F}{A} = \frac{u}{2} \int_0^\infty \left(|\psi_0|^4 - |\psi|^4 \right) \mathrm{d}x .$$

In the following, we refer to this energy difference per area as *surface energy* $\sigma := \Delta F/A$. After using the solution (4.35), we obtain an analytically solvable integral for σ,

$$\sigma = \frac{u}{2} \psi_0^4 \int_0^\infty \left[1 - \tanh^4 \left(\frac{x}{\xi\sqrt{2}} \right) \right] \mathrm{d}x$$

$$= \frac{u}{2} \psi_0^4 \xi\sqrt{2} \int_0^\infty \left(1 - \tanh^4 t \right) \mathrm{d}t = \frac{\sqrt{8}}{3} \xi u \psi_0^4 , \qquad (4.43)$$

where we have used the standard integral

$$\int_0^\infty \left(1 - \tanh^4 t \right) \mathrm{d}t = 4/3 .$$

We see that the surface energy is always positive and proportional to the coherence length. Thus, according to definition (4.42), the free energy of a superconducting material with a finite coherence length is higher than the energy of a material with a constant ψ. In other words, the presence of an interface to a region with normal conductivity is energetically costly. This situation may change if an external magnetic field is applied. As is shown in the next section, under certain conditions the surface energy may become negative, leading to preferred formation of normal conducting domains.

Before taking into account magnetic fields, let us finally express the phenomenological parameter ψ_0 in (4.43) by a measurable quantity. In the next chapter, we will show that the critical magnetic field H_c is related to r and u by the following equation,

$$\frac{|r|^2}{u} = \frac{H_c^2}{4\pi} .$$

On the other hand, due to $|\psi_0|^2 = |r|/u$, we may write

$$|\psi_0|^4 u = \frac{|r|^2}{u} = \frac{H_c^2}{4\pi}.$$

Inserting this result into (4.43), the surface energy reads

$$\sigma = \frac{\sqrt{32}}{3} \frac{H_c^2}{8\pi} \xi \approx 1.89 \, |\Delta f| \, \xi, \tag{4.44}$$

where $|\Delta f| = H_c^2/(8\pi)$ is the field energy density due to the Meissner effect. We see that the surface energy is proportional to the coherence length and a fraction of the field energy.

4.3 – Superconductor in a magnetic field

In this section, we study the general case of the Ginzburg-Landau theory by taking into account external magnetic fields. Starting point is again the free energy, which we rewrite such that the coupling to external fields is taken into account. Thereby, we start again with the basic assumption that the superconducting order parameter $\psi(\vec{r})$ is considered as the wave function of the superconducting many-particle system. Within the quantum theory the magnetic field is taken into account using an extension of the momentum operator,

$$\frac{\hbar}{i} \vec{\nabla} \rightarrow \frac{\hbar}{i} \vec{\nabla} - \frac{q^*}{c} \vec{A}, \tag{4.45}$$

where the vector potential $\vec{A}$ is connected to the magnetic induction $\vec{B}$ via the known relation

$$\vec{B} = \vec{\nabla} \times \vec{A}. \tag{4.46}$$

The constant q^* is the charge of the superconducting particle.

As it was introduced in section 3.6, the free energy of a material in the presence of a magnetic field is a functional with respect to the magnetic induction $\vec{B}$, and it depends explicitly on T. It must have the form (3.21). In the superconducting state, the free energy additionally obtains a functional dependence on ψ, which occurs in the magnetization part F_M. Thus, according to (3.21) we can write

$$F[\vec{B},T,\psi] = \frac{1}{8\pi}\int_{\mathscr{V}} d^3r\,\vec{B}^2(\vec{r}) + F_M[\vec{B},T,\psi]. \qquad (4.47)$$

The explicit expression for F_M can be easily found from (4.16). Here we only have to replace the momentum operator $(\hbar/i)\,\vec{\nabla}$ with the extension (4.45). We obtain

$$\begin{aligned}
F_M[\vec{B},T,\psi] = {}& F_{M,n}[\vec{B},T] \\
& + \int_{\mathscr{V}} d^3r\left[\frac{1}{2m^*}\left|\left(\frac{\hbar}{i}\vec{\nabla} - \frac{q^*}{c}\vec{A}(\vec{r})\right)\psi(\vec{r})\right|^2\right. \\
& \left. + r\,|\psi(\vec{r})|^2 + \frac{u}{2}|\psi(\vec{r})|^4\right],
\end{aligned} \qquad (4.48)$$

where $F_{M,n}[\vec{B},T]$ is the magnetization part of the free energy in the normal state. Note that the coupling of the superconducting order parameter to the magnetic field takes place via the vector potential $\vec{A}$ in the second term in (4.48). The vector potential, in turn, is connected to the magnetic induction $\vec{B}$ via the relation (4.46). Therefore, in the superconducting state, an additional magnetization is introduced which contributes to the functional dependence of the total free energy on $\vec{B}$.

Due to the occurrence of the vector potential in the magnetization term (4.48), it is easier for the following considerations to treat the free energy with $\vec{A}$ (instead of $\vec{B}$) as a natural variable. From (4.47) and (4.48) with the replacement of $\vec{B}$ with $\vec{A}$ using (4.46) we obtain the following expression for the free energy, which we will use as the starting point for the theory,

$$F[\vec{A},T,\psi] = F_{M,n}[\vec{A},T] + \frac{1}{8\pi}\int_{\mathscr{V}} \mathrm{d}^3r\,(\vec{\nabla}\times\vec{A})^2$$

$$+\int_{\mathscr{V}} \mathrm{d}^3r\left[\frac{1}{2m^*}\left|\left(\frac{\hbar}{i}\vec{\nabla}-\frac{q^*}{c}\vec{A}(\vec{r})\right)\psi(\vec{r})\right|^2\right.$$

$$\left. +r\,|\psi(\vec{r})|^2 + \frac{u}{2}|\psi(\vec{r})|^4\right]. \tag{4.49}$$

This expression represents the Landau potential for a superconductor in the magnetic field.

4.3.1 – Gauge invariance

Before we find out the basic equations of the Ginzburg-Landau theory, it is useful to check the gauge symmetry of the Landau potential. If the Landau potential were invariant with respect to a gauge transformation, which will be defined next, this would mean that the Ginzburg-Landau equations would also apply to the transformed variables. The gauge transformation can then be arranged in such a way that the transformed variables have a particularly simple and adapted form to the needs. Nevertheless, these new variables fulfill the original Ginzburg-Landau equations. This can be very helpful in solving these equations. For example, a specific gauge transformation can make the order parameter real. This simplifies the handling of the order parameter and the solution of the differential equations, but leaves any physical quantity calculated from it unchanged. Therefore, gauge transformations are generally of great importance in the practical implementation of the solution of equations of motion, such as in the electrodynamics the Maxwell equations.

In our case of a superconductor, the gauge freedom is applied as follows. At first, we show that the functional $F[\vec{A},T,\psi]$ is invariant under a local gauge transformation of the form

$$\psi(\vec{r}) \rightarrow e^{i\alpha(\vec{r})}\psi(\vec{r}), \qquad \vec{A}(\vec{r}) \rightarrow \vec{A}(\vec{r}) + \frac{\hbar c}{q^*}\vec{\nabla}\alpha(\vec{r}),$$

where $\alpha(\vec{r})$ is an arbitrary scalar field. Note that the second gauge transformation for the vector potential $\vec{A}$ keeps the magnetic induction $\vec{B} = \vec{\nabla} \times \vec{A}$ invariant since for any gradient field the relationship $\vec{\nabla} \times (\vec{\nabla}\alpha) = 0$ is valid.

Now we check whether the Landau potential (4.49) actually remains invariant with respect to the above transformations. At first, we introduce the transformed fields,

$$\psi'(\vec{r}) = e^{i\alpha(\vec{r})}\psi(\vec{r}), \tag{4.50}$$

$$\vec{A}'(\vec{r}) = \vec{A}(\vec{r}) + \frac{\hbar c}{q^*}\vec{\nabla}\alpha(\vec{r}), \tag{4.51}$$

and then we show that the expression (4.49) keeps invariant if we replace ψ and $\vec{A}$ with ψ' and $\vec{A}'$. While transformation (4.50) changes the *phase* of the complex function $\psi(\vec{r})$, its absolute value remains unchanged, $|\psi'| = |\psi|$. Thus, we immediately see that the integrals over the fields $|\psi|^2$ and $|\psi|^4$ in (4.49) are not changed. One can also immediately see that the field energy term

$$\frac{1}{8\pi}\int_{\mathscr{V}} d^3r\,\vec{B}^2(\vec{r})$$

does not change under transformation (4.51), because as already mentioned, a gradient field does not affect a physical field.

Finally, only the term with the momentum operator in (4.49) remains to be examined. We insert the transformed fields (4.50) and (4.51) into the extended momentum operator applied to the order parameter and obtain

$$\left(\frac{\hbar}{i}\vec{\nabla} - \frac{q^*}{c}\vec{A}'\right)\psi' = \left(\frac{\hbar}{i}\vec{\nabla} - \frac{q^*}{c}\vec{A} - \hbar[\vec{\nabla}\alpha]\right)e^{i\alpha}\psi,$$

where $[\vec{\nabla}\alpha]$ means that the Nabla operator acts solely on the function $\alpha(\vec{r})$. Then, we shall apply the known product and chain rules for differential operators to regain the original form of the extended momentum operator. We find

$$\left(\frac{\hbar}{i}\vec{\nabla} - \frac{q^*}{c}\vec{A}'\right)\psi' = \frac{\hbar}{i}e^{i\alpha}\left(\psi i\vec{\nabla}\alpha + \vec{\nabla}\psi\right)$$

$$- \left(\frac{q^*}{c}\vec{A} + \hbar[\vec{\nabla}\alpha]\right)e^{i\alpha}\psi = \frac{\hbar}{i}e^{i\alpha}\vec{\nabla}\psi - \frac{q^*}{c}\vec{A}e^{i\alpha}\psi$$

$$= e^{i\alpha}\left(\frac{\hbar}{i}\vec{\nabla} - \frac{q^*}{c}\vec{A}\right)\psi. \tag{4.52}$$

The transformed momentum operator applied to the transformed order parameter field has the original form up to the phase factor $e^{i\alpha}$. From (4.52) we see that the absolute value is not changed under the gauge transformation,

$$\left|\left(\frac{\hbar}{i}\vec{\nabla} - \frac{q^*}{c}\vec{A}'\right)\psi'\right|^2 =$$

$$\left[\left(\frac{\hbar}{i}\vec{\nabla} - \frac{q^*}{c}\vec{A}\right)\psi\right]^* e^{-i\alpha} \cdot e^{i\alpha} \left[\left(\frac{\hbar}{i}\vec{\nabla} - \frac{q^*}{c}\vec{A}\right)\psi\right]$$

$$= \left|\left(\frac{\hbar}{i}\vec{\nabla} - \frac{q^*}{c}\vec{A}\right)\psi\right|^2,$$

which proves that the magnetization part of the free energy (4.49) is invariant under the gauge transformations (4.50) and (4.51).

In conclusion, while the order parameter function as well as the vector potential depend on the particular gauge, physical quantities, such as the free energy or the magnetic field, remain unchanged. This property will help us later to solve the Ginzburg-Landau equations.

4.3.2 – Set of basic equations

Let us now derive the Ginzburg-Landau equations in the presence of a magnetic field. The resulting set of differential equations determines the equilibrium state of the superconducting order parameter. To find these equations, we carry out the minimization of the free energy as already discussed in sections 4.1 and 4.2. In the presence of a magnetic field, the free energy is formally a functional with respect to five functions: $\psi(\vec{r})$, its complex conjugate $\psi^*(\vec{r})$, and the three components of $\vec{A}(\vec{r})$. The minimization procedure results in the two differential equations for ψ and ψ^*. In addition, the functional derivative of the free energy with respect to the vector potential, which is related via equation (3.16) to the external magnetic field, can be used to derive three further differential equations which determine the three components of the magnetic induction inside the superconductor. Thus, in total, we will arrive at five Ginzburg-Landau equations.

The most convenient way of minimization is to use the concept of the functional derivative, as introduced in the section 3.3. We start with the minimization with respect to ψ^*. As already described for the above case without magnetic field, the volume integral in the free energy has to be written such that ψ^* appears explicitly and that no differential operator acts on ψ^*. So we start with the following description of free energy based on an energy density,

$$F = F_{M,n} + \int_{\mathcal{V}} \mathrm{d}^3 r f(\vec{A}(\vec{r}), T, \psi(\vec{r}), \psi^*(\vec{r})), \qquad (4.53)$$

where the energy density f is found from (4.49).

Next, we rewrite the function f in a way that the functional derivative with respect to ψ^* can be applied. At first, we expand the scalar product which appears as the term with the momentum operator in (4.49),

$$\left| \left(\frac{\hbar}{i} \vec{\nabla} - \frac{q^*}{c} \vec{A} \right) \psi \right|^2 =$$

$$\left(\frac{\hbar}{i} \vec{\nabla}\psi - \frac{q^*}{c} \vec{A}\psi \right) \cdot \left(-\frac{\hbar}{i} \vec{\nabla}\psi^* - \frac{q^*}{c} \vec{A}\psi^* \right)$$

$$= \hbar^2 [\vec{\nabla}\psi] \cdot [\vec{\nabla}\psi^*] - \frac{\hbar q^*}{ic} \left(\psi^* \vec{A} \cdot \vec{\nabla}\psi \right.$$

$$\left. - \psi \vec{A} \cdot \vec{\nabla}\psi^* \right) + \frac{q^{*2}}{c^2} \vec{A}^2 |\psi|^2 . \qquad (4.54)$$

The volume integral over the terms containing $\vec{\nabla}\psi^*$ can be reformulated as in (4.22) using the identity (4.20) and the Gauss theorem which is applied to a closed surface outside the superconducting volume. The relevant terms are

$$\int_{\mathcal{V}} d^3r\,\psi\vec{A}\cdot\vec{\nabla}\psi^* = -\int_{\mathcal{V}} d^3r\,\psi^*\vec{\nabla}\cdot(\psi\vec{A}), \tag{4.55}$$

$$\int_{\mathcal{V}} d^3r\,[\vec{\nabla}\psi]\cdot[\vec{\nabla}\psi^*] = -\int_{\mathcal{V}} d^3r\,\psi^*\vec{\nabla}^2\psi. \tag{4.56}$$

Now we use the expansion (4.54) to replace the corresponding term in (4.49). Taking into account (4.55) and (4.56) we find an expression for the free energy where ψ^* appears explicitly as a pre-factor inside the volume integral,

$$
\begin{aligned}
F[\vec{A},T,\psi] = F_{M,n}[\vec{A},T] + \int_{\mathcal{V}} d^3r\Bigg[&-\frac{\hbar^2}{2m^*}\psi^*\vec{\nabla}^2\psi \\
&-\frac{\hbar q^*}{ic\,2m^*}\left(\psi^*\vec{A}\cdot\vec{\nabla}\psi + \psi^*\vec{\nabla}\cdot(\psi\vec{A})\right) \\
&+\frac{1}{2m^*}\left(\frac{q^*}{c}\right)^2\vec{A}^2|\psi|^2 + r|\psi|^2 \\
&+\frac{u}{2}|\psi|^4 + \frac{1}{8\pi}(\vec{\nabla}\times\vec{A})^2\Bigg].
\end{aligned}
\tag{4.57}
$$

The energy density inside the volume integral corresponds to the specific function f in (4.53). It is given by

$$
\begin{aligned}
f(\vec{A},T,\psi,\psi^*) = &-\frac{\hbar^2}{2m^*}\psi^*\vec{\nabla}^2\psi \\
&-\frac{\hbar q^*}{ic\,2m^*}\left(\psi^*\vec{A}\cdot\vec{\nabla}\psi + \psi^*\vec{\nabla}\cdot(\psi\vec{A})\right) \\
&+\left(\frac{q^{*2}}{2m^*c^2}\vec{A}^2 + r\right)\psi\psi^* \\
&+\frac{u}{2}\psi^2(\psi^*)^2 + \frac{1}{8\pi}(\vec{\nabla}\times\vec{A})^2.
\end{aligned}
\tag{4.58}
$$

Following the ideas of section 3.3, the functional derivative with respect to ψ^* is given by the partial derivative $\partial f/\partial \psi^*$ of the function (4.58) considered at position $\vec{r}$,

$$\frac{\delta F}{\delta \psi^*(\vec{r})} = \frac{\partial f(\vec{A},T,\psi,\psi^*)}{\partial \psi^*}(\vec{r}).$$

To calculate the functional derivative, we therefore only have to take the expression (4.58) and form the partial derivative with respect to ψ^*. We obtain

$$\frac{\delta F}{\delta \psi^*(\vec{r})} = -\frac{\hbar^2}{2m^*}\vec{\nabla}^2\psi - \frac{\hbar q^*}{ic\,2m^*}\left(\vec{A}\cdot\vec{\nabla}\psi + \vec{\nabla}\cdot(\psi\vec{A})\right)$$
$$+ \left(\frac{q^{*2}}{2m^*c^2}\vec{A}^2 + r\right)\psi + u|\psi|^2\psi,$$

where in this equation all fields refer to position $\vec{r}$. Finally, we factor out $(-\hbar^2/2m^*)$ and reintroduce the extended momentum operator,

$$\frac{\delta F}{\delta \psi^*(\vec{r})} = -\frac{\hbar^2}{2m^*}\left(\vec{\nabla}^2\psi + \frac{q^*}{\hbar ic}\left[\vec{A}\cdot\vec{\nabla}\psi + \vec{\nabla}\cdot(\psi\vec{A})\right]\right.$$
$$\left. - \frac{q^{*2}}{\hbar^2 c^2}\vec{A}^2\psi\right) + r\psi + u|\psi|^2\psi.$$

The minimization condition claims that the functional derivative vanishes. Thus, the final result for the first Ginzburg-Landau equation in the presence of a magnetic field reads

$$-\frac{\hbar^2}{2m^*}\left(\vec{\nabla} - \frac{iq^*}{\hbar c}\vec{A}\right)^2\psi + r\psi + u|\psi|^2\psi = 0. \qquad (4.59)$$

Alternatively, introducing again the Ginzburg-Landau coherence length $\xi = \hbar/\sqrt{2m^*|r|}$ and taking into account $r < 0$, we may write the above Ginzburg-Landau equation (4.59) as follows,

$$-\xi^2 \left(\vec{\nabla} - \frac{iq^*}{\hbar c}\vec{A} \right)^2 \psi - \psi + \frac{u}{|r|}|\psi|^2 \psi = 0. \qquad (4.60)$$

Again, ξ acts as a characteristic length scale in which the order parameter changes. The minimization of F with respect to ψ (not shown here) leads to the complex conjugate of equation (4.60).

Next, we turn to the evaluation of equation (3.16),

$$\vec{\nabla} \times \vec{H} = 4\pi \frac{\delta F}{\delta \vec{A}}, \qquad (4.61)$$

which leads us to three further equations which add to the set of basic equations of the Ginzburg-Landau theory. It is convenient to start from a decomposition of F according to (4.49) into two parts,

$$F = F_M + F_{em}, \qquad (4.62)$$

where F_M is the magnetization part and F_{em} is the field part of the free energy. It is defined as follows

$$F_{em} = \frac{1}{8\pi} \int_{\mathcal{V}} \mathrm{d}^3 r \left(\vec{\nabla} \times \vec{A}(\vec{r}) \right)^2. \qquad (4.63)$$

According to equation (4.61), the functional derivative with respect to the component A_j of the vector potential leads us to the j component of $\vec{\nabla} \times \vec{H}$. Thus, with the decomposition (4.62) we obtain

$$\frac{1}{4\pi}(\vec{\nabla} \times \vec{H})_j = \frac{\delta F}{\delta A_j} = \frac{\delta F_M}{\delta A_j} + \frac{\delta F_{em}}{\delta A_j}. \qquad (4.64)$$

Let us start with the functional derivative of F_M. We use the extended formulation of (4.57), and for the functional derivative we have to take into account only the terms containing $\vec{A}$,

$$\frac{\delta F_M}{\delta A_j} = \frac{\delta}{\delta A_j} \int_{\mathcal{V}} \mathrm{d}^3 r \left[-\frac{\hbar q^*}{ic\,2m^*} \left(\psi^* \vec{A} \cdot \vec{\nabla}\psi + \psi^* \vec{\nabla} \cdot (\psi\vec{A}) \right) \right.$$
$$\left. + \frac{1}{2m^*} \left(\frac{q^*}{c} \right)^2 \vec{A}^2 |\psi|^2 \right] + \frac{\delta F_{M,n}}{\delta A_j} . \qquad (4.65)$$

For simplification, we will always omit the last term, $\delta F_{M,n}/\delta A_j$ from now on. This corresponds to the approximation that the material cannot be magnetized in the normal state which is a well fulfilled property for most metals.

To easily calculate the functional derivative of the volume integral in (4.65), we rewrite the second term of the volume integral using again the rule (4.20) and the Gauss theorem,

$$\int_{\mathcal{V}} \mathrm{d}^3 r\, \psi^* \vec{\nabla} \cdot (\psi\vec{A}) = -\int_{\mathcal{V}} \mathrm{d}^3 r\, \psi \vec{A} \cdot \vec{\nabla}\psi^* .$$

With the help of this relation, we can evaluate the functional derivative as usual by computing the partial derivative of the integrand with respect to the component A_j and considering the resulting expression at position vector $\vec{r}$,

$$\frac{\delta F_M}{\delta A_j} = -\frac{\hbar q^*}{ic\,2m^*} \left(\psi^* \frac{\partial \psi}{\partial x_j} - \psi \frac{\partial \psi^*}{\partial x_j} \right) + \frac{q^{*2}}{m^* c^2} A_j |\psi|^2 ,$$

or in vector notation

$$\frac{\delta F_M}{\delta \vec{A}} = -\frac{\hbar q^*}{ic\,2m^*} \left(\psi^* \vec{\nabla}\psi - \psi \vec{\nabla}\psi^* \right) + \frac{q^{*2}}{m^* c^2} |\psi|^2 \vec{A} . \quad (4.66)$$

Finally, the variation of the functional F_{em} remains to be calculated. Since F_{em} only depends on $\vec{A}$, a variation of F_{em} can be written in terms of a variation of $\vec{A}$,

$$\delta F_{em} = \frac{1}{4\pi} \int_{\mathcal{V}} d^3r (\vec{\nabla} \times \vec{A}) \cdot (\vec{\nabla} \times \delta \vec{A})$$

$$= \frac{1}{4\pi} \int_{\mathcal{V}} d^3r \, \vec{B} \cdot (\vec{\nabla} \times \delta \vec{A}),$$

with $\vec{B} = \vec{\nabla} \times \vec{A}$. Using again the property

$$\vec{\nabla} \cdot (\vec{a} \times \vec{b}) = \vec{b} \cdot \vec{\nabla} \times \vec{a} - \vec{a} \cdot \vec{\nabla} \times \vec{b},$$

which is valid for arbitrary vector fields $\vec{a}$ and $\vec{b}$, we rewrite the volume integral in δF_{em} as follows,

$$\delta F_{em} = \frac{1}{4\pi} \int_{\mathcal{V}} d^3r \, \vec{\nabla} \cdot (\delta \vec{A} \times \vec{B}) + \frac{1}{4\pi} \int_{\mathcal{V}} d^3r \, \delta \vec{A} \cdot (\vec{\nabla} \times \vec{B}).$$

$$(4.67)$$

In the first term the Gauss theorem can be used again to replace the volume integral by the surface integral over a closed surface $\mathcal{S}(\mathcal{V})$ of the superconducting volume $\mathcal{V}$. Therefore, this term reads

$$\int_{\mathcal{V}} d^3r \, \vec{\nabla} \cdot (\delta \vec{A} \times \vec{B}) = \oint_{\mathcal{S}(\mathcal{V})} d\vec{f} \cdot (\delta \vec{A} \times \vec{B}).$$

Per definition, the variation $\delta \vec{A}$ is set to zero in all places on the surface $\mathcal{S}(\mathcal{V})$. Consequently, the first term of (4.67) vanishes and the variation is fully determined by the second term. Thus, the functional derivative can be written down directly,

$$\frac{\delta F_{em}}{\delta \vec{A}(\vec{r})} = \frac{1}{4\pi} \operatorname{rot} \vec{B}(\vec{r}).$$

$$(4.68)$$

Finally, using equation (4.64) with the expressions (4.66) and (4.68), we arrive at the third Ginzburg-Landau equation,

$$\boxed{\begin{aligned} \operatorname{rot}\vec{H} = \operatorname{rot}\vec{B} + \\ \frac{4\pi}{c}\left[-\frac{\hbar q^*}{i2m^*}\left(\psi^* \vec{\nabla}\psi - \psi \vec{\nabla}\psi^* \right) + \frac{q^{*2}}{m^*c}|\psi|^2 \vec{A} \right]. \end{aligned}}$$

$$(4.69)$$

In summary, we have a set of three non-linear differential equations:

$$0 = -\xi^2 \left(\vec{\nabla} - \frac{iq^*}{\hbar c}\vec{A} \right)^2 \psi - \psi + \frac{u}{|r|}|\psi|^2\psi, \qquad (4.70)$$

$$0 = -\xi^2 \left(\vec{\nabla} + \frac{iq^*}{\hbar c}\vec{A} \right)^2 \psi^* - \psi^* + \frac{u}{|r|}|\psi|^2\psi^*, \qquad (4.71)$$

$$\mathrm{rot}\,\vec{H} = \frac{4\pi}{c}\left[-\frac{\hbar q^*}{i2m^*}\left(\psi^*\vec{\nabla}\psi - \psi\vec{\nabla}\psi^* \right) + \frac{q^{*2}}{m^*c}|\psi|^2\vec{A} \right]$$
$$+ \vec{\nabla}\times(\vec{\nabla}\times\vec{A}). \qquad (4.72)$$

It can be considered as the basic set of equations of the Ginzburg-Landau theory. For the given external magnetic field $\vec{H}$ and boundary conditions, the system of equations (4.70)-(4.72) determines the fields $\psi(\vec{r})$, $\psi^*(\vec{r})$, and $\vec{A}(\vec{r})$. In general, due to the non-linear structure of the differential equations, numerical methods have to be employed to solve them. This results in the fields of the complex order parameter $\psi(\vec{r}) = |\psi(\vec{r})|e^{i\Theta(\vec{r})}$ and the vector potential $\vec{A}(\vec{r})$. From these calculated fields, measurable physical quantities can then be determined. Examples are the magnetic induction $\vec{B} = \vec{\nabla}\times\vec{A}$ or the density of the superconducting particles $n_s^* = |\psi|^2$.

As is shown further below, the phenomenological parameters $u(T)$ and $r(T)$ can be expressed in terms of the measurable quantities $\lambda_L(T)$ and $H_c(T)$. Furthermore, it should be noted that an externally created magnetic field $\vec{H}$ enters the theory via the left side of equation (4.72) and also via the boundary condition of the vector potential $\vec{A}$. The solution of the above system of equations (4.70)-(4.72) obtained under these conditions then results in the vector potential and order parameter inside the superconductor as a reaction to the external field. Such a consideration plays an important role, in particular, if one is interested in the specific external field at which the superconducting state collapses (critical magnetic field). This will be discussed in more detail below in the next chapter.

4.4 – Property of superconductivity

In this section, we want to show how the property of superconductivity enters the Ginzburg-Landau theory. This should be done in two ways. First, we calculate the electric current density within the framework of the Ginzburg-Landau formalism and show that an electric current can flow in the material without having to create an external electric field. This property of an infinitely large electrical conductivity defines a superconductor. Secondly, we will show that the London theory is completely included in the Ginzburg-Landau theory. This is achieved by deriving the second of London's equations from the presented formalism.

4.4.1 – Supercurrent and superfluid velocity

At first, let us elaborate how the current density $j_s(\vec{r})$ of superconducting particles arises within the Ginzburg-Landau formalism. This quantity was already introduced in chapter 2. In the following, we will use the abbreviated name 'supercurrent'. We show how the supercurrent is calculated within the framework of the Ginzburg-Landau formalism and that the phase $\Theta(\vec{r})$ of the order parameter plays an important role in this.

Applying the curl to the general relation (3.17) leads us to the equation $\mathrm{rot}\,\vec{H} = \mathrm{rot}\,\vec{B} - 4\pi\,\mathrm{rot}\,\vec{M}$ which can be easily compared with the above result (4.69). We obtain an expression for the curl of the magnetization $\vec{M}$ of the entire system,

$$c\,\mathrm{rot}\vec{M} = \frac{\hbar q^*}{i2m^*}\left(\psi^*\vec{\nabla}\psi - \psi\vec{\nabla}\psi^*\right) - \frac{q^{*2}}{m^*c}|\psi|^2\vec{A}. \quad (4.73)$$

From electrodynamics it is known that a distribution of magnetic dipoles with the magnetization $\vec{M}$ generates the current density $\vec{j} = c\,\mathrm{rot}\vec{M}$. Since the magnetization is fully caused by currents in the superconducting condensate, the right side of (4.73) can be interpreted as the current density of the superconducting particles which we call supercurrent. It is given by

$$\vec{j}_s = \frac{\hbar q^*}{i2m^*}\left(\psi^*\vec{\nabla}\psi - \psi\vec{\nabla}\psi^*\right) - \frac{q^{*2}}{m^*c}|\psi|^2\vec{A}.\qquad(4.74)$$

Note that the supercurrent $\vec{j}_s$ has the same structure as the current of a general wave function in a magnetic field (compare section 2.3). A detailed discussion of the relation to the London theory is given below.

Let us now find an expression for the velocity of the superconducting particles. For this purpose, we use the complex order parameter function in the form $\psi(\vec{r}) = |\psi(\vec{r})|e^{i\Theta(\vec{r})}$ to write the expression (4.74) in terms of the absolute value $|\psi(\vec{r})|$ and the phase $\Theta(\vec{r})$. In general, both fields depend on $\vec{r}$. The gradient of ψ reads

$$\vec{\nabla}\psi = e^{i\Theta}\vec{\nabla}|\psi| + |\psi|e^{i\Theta}i\vec{\nabla}\Theta = e^{i\Theta}\vec{\nabla}|\psi| + i\psi\vec{\nabla}\Theta.$$
$$(4.75)$$

A very useful formula for later discussions is obtained if one replaces in (4.74) all gradient fields by expression (4.75). At first, one obtains

$$\vec{j}_s(\vec{r}) = \frac{\hbar q^*}{i2m^*}\left[\psi^*\left(e^{i\Theta}\vec{\nabla}|\psi| + i\psi\vec{\nabla}\Theta\right)\right.$$
$$\left. - \psi\left(e^{-i\Theta}\vec{\nabla}|\psi| - i\psi^*\vec{\nabla}\Theta\right)\right] - \frac{q^{*2}}{m^*c}|\psi|^2\vec{A}.\qquad(4.76)$$

Due to the property $\psi^*e^{i\Theta} = |\psi| = \psi e^{-i\Theta}$ the two terms proportional to $\vec{\nabla}|\psi|$ in (4.76) cancel each other. Thus, only the spatial variation of the phase Θ contributes to the supercurrent. We obtain

$$\vec{j}_s = \frac{\hbar q^*}{i2m^*}2i|\psi|^2\vec{\nabla}\Theta - \frac{q^{*2}}{m^*c}|\psi|^2\vec{A}$$
$$= \frac{\hbar q^*}{m^*}|\psi|^2\left(\vec{\nabla}\Theta - \frac{q^*}{\hbar c}\vec{A}\right).\qquad(4.77)$$

The density $|\psi|^2 = n_s^*$ is interpreted as the density of the superconducting particles (Cooper pairs). Note that the particle density

94

$|\psi|^2$ generally depends on the position vector, i. e. a scalar field $|\psi|^2 = n_s^* = n_s^*(\vec{r})$ has to be considered in general. This becomes particularly important when the superconducting state near an interface to a normal conducting metal is of interest. Thus, the supercurrent $\vec{j}_s$ takes a generalized form of (2.1) which was already introduced within the London theory. We write

$$\vec{j}_s(\vec{r}) = q^* n_s^*(\vec{r})\, \vec{v}_s(\vec{r}) \tag{4.78}$$

with the superfluid velocity

$$\vec{v}_s(\vec{r}) = \frac{\hbar}{m^*}\left[\vec{\nabla}\Theta(\vec{r}) - \frac{q^*}{\hbar c}\vec{A}(\vec{r})\right]. \tag{4.79}$$

Using the expressions (4.78) and (4.79) we can finally calculate the supercurrent from the fields $\psi(\vec{r})$ and $\vec{A}(\vec{r})$ which are obtained from the solution of the Ginzburg-Landau equations.

It is seen that the supercurrent is partially determined by the gradient of the order parameter phase. This means that as soon as the phase of the order parameter varies spatially, a supercurrent flows (even without a magnetic field, $\vec{A} = 0$). Note that no electric field is required for this. Consequently, it is the property of the phase of the order parameter that describes the superconductivity.

4.4.2 – Relation to the London theory

Another way to prove that the order parameter in the Ginzburg-Landau theory describes superconductivity is to show that the theory contains the London equations. This is to be done in the following with the help of a special gauge transformation.

At first, let us show that the superfluid velocity does not change if the fields $\Theta(\vec{r})$ and $\vec{A}(\vec{r})$ are changed to other values $\Theta'(\vec{r})$ and $\vec{A}'(\vec{r})$ by a local gauge transformation. With this knowledge, we then apply a particular gauge which leads us to the London equations. Starting with the superfluid velocity in terms of transformed fields,

$$\vec{v}_s'(\vec{r}) = \frac{\hbar}{m^*}\left[\vec{\nabla}\Theta'(\vec{r}) - \frac{q^*}{\hbar c}\vec{A}'(\vec{r})\right],$$

we replace the transformed fields by the original fields using (4.50) and (4.51). The first aim is to show that $\vec{v}_s' = \vec{v}_s$. According to (4.50) the relation between the transformed phase and the original phase is $\Theta' = \Theta + \alpha$. The corresponding relationship for the vector potential is directly given by (4.51). Using all these relationships step by step we find

$$\begin{aligned}
\vec{v}_s' &= \frac{\hbar}{m^*}\left[\vec{\nabla}\Theta' - \frac{q^*}{\hbar c}\vec{A}'\right] \\
&= \frac{\hbar}{m^*}\left[\vec{\nabla}(\Theta + \alpha) - \frac{q^*}{\hbar c}\left(\vec{A} + \frac{\hbar c}{q^*}\vec{\nabla}\alpha\right)\right] \\
&= \frac{\hbar}{m^*}\left[\vec{\nabla}\Theta + \vec{\nabla}\alpha - \frac{q^*}{\hbar c}\vec{A} - \vec{\nabla}\alpha\right] \\
&= \frac{\hbar}{m^*}\left[\vec{\nabla}\Theta - \frac{q^*}{\hbar c}\vec{A}\right] = \vec{v}_s.
\end{aligned}$$

The superfluid velocity does not change under a local gauge transformation. Due to $n_s^* = |\psi|^2$ it is easily seen from (4.78) that the supercurrent also remains unchanged.

We have shown that all physical quantities remain invariant with respect to any gauge transformation of the form (4.50) and (4.51). Applying the special (global) gauge transformation which is constructed in such a way that the phase vanishes in the whole volume ($\Theta_L = 0$, London gauge), we see from equations (4.78) and (4.79) that in this gauge the formula of the current density is

$$\vec{j}_s(\vec{r}) = -\frac{q^{*2}}{m^*c}n_s^*(\vec{r})\vec{A}_L(\vec{r}), \tag{4.80}$$

where $\vec{A}_L(\vec{r})$ is the vector potential in the London gauge. Note that the particle density $n_s^*(\vec{r})$ generally varies in space.

According to the section 2.1, the London theory starts from the approximation that an incompressible liquid of electrons carries the superconductivity. Within the general Ginzburg-Landau formalism, this approximation means that the particle density strictly keeps unchanged. Therefore, the London theory is valid in the special case that the particle density can approximately be set to a constant, $n_s^*(\vec{r}) \approx n_s^*$.

Furthermore, it is assumed that two 'superconducting' electrons form a Cooper pair leading to the relations $q^* = -2e$, $n_s = 2n_s^*$, and $m = m^*/2$, where $(-e)$ is the charge and m the mass of one conduction electron (as considered within the London theory). Replacing the Cooper pair quantities in (4.80) with the constant electron density and the other quantities referring to one electron we find

$$\vec{j}_s(\vec{r}) = -\frac{n_s e^2}{mc}\vec{A}_L(\vec{r}), \tag{4.81}$$

which corresponds to the result (2.22) of the London theory.

The second of London's basic equations (2.11), which determines together with the Ampere's law the magnetic field and current density inside the superconductor, is obtained by forming the curl on both sides of equation (4.81),

$$\vec{\nabla} \times \vec{j}_s(\vec{r}) = -\frac{n_s e^2}{mc}\vec{\nabla} \times \vec{A}_L(\vec{r}) = -\frac{n_s e^2}{mc}\vec{B}(\vec{r}). \tag{4.82}$$

We see that the London theory is fully consistent with the Ginzburg-Landau theory. The London theory is obtained in the special case of a constant particle density.

In the London theory, we have shown that the property of superconductivity in particular leads to the special form of the second London equation. Since the Ginzburg-Landau theory obviously provides the second London equation without further assumptions, it is shown that the complex order parameter ψ describes the property superconductivity.

4.4.3 – Supercurrent versus order parameter

As the perhaps simplest application of the Ginzburg-Landau theory, let us consider a superconductor where both the particle density and the current density are constant in space. According to the previous considerations, the London theory is valid in this case. Thus, it follows from the field equation (2.8) that in the case of a constant current density $\vec{j}_s$, the magnetic induction is zero, i. e. $\vec{B} = 0$.

Instead of solving the Ginzburg-Landau equations, as one would proceed in the general case, it proves to be much easier for this example to express the free energy in terms of the constant parameter $|\psi|$ and then minimize it directly with respect to a variation of $|\psi|$. The aim of this procedure is to find $|\psi|$ as a function of the supercurrent or the superfluid velocity. Of particular interest here is the value of the critical current, at which the superconducting state disappears.

As fixed in our starting point, we assume that the coherence length is large enough, that both $|\psi|$ and $v_s = |\vec{v}_s|$ can be considered as constant in space. The magnetic induction is zero. We start directly from the general expression (4.48) of the Ginzburg-Landau free energy with $\vec{A} = 0$,

$$
F_M = F_{M,n} + \int_{\mathcal{V}} d^3r \left[\frac{1}{2m^*} \left| \frac{\hbar}{i} \vec{\nabla} \psi(\vec{r}) \right|^2 \right.
$$
$$
\left. + r|\psi(\vec{r})|^2 + \frac{u}{2}|\psi(\vec{r})|^4 \right],
\tag{4.83}
$$

instead of solving the equations (4.70)-(4.72). Note that $F = F_M$.

Our first initial assumption is $|\psi(\vec{r})| = |\psi|$, i. e. the absolute value of the order parameter is constant in space. Note, however, that the phase is not constant. This condition allows us to rewrite the vector $(\hbar/i)\vec{\nabla}\psi(\vec{r})$ in (4.83) in terms of $\vec{v}_s(\vec{r})$, which is achieved by simply inserting ψ with a constant absolute value,

$$\frac{\hbar}{i}\,\vec{\nabla}\,|\psi|\,e^{i\Theta(\vec{r})} = |\psi|\,\hbar e^{i\Theta(\vec{r})}\,\vec{\nabla}\,\Theta(\vec{r})$$

$$= m^*\,|\psi|\,e^{i\Theta(\vec{r})}\,\frac{\hbar}{m^*}\,\vec{\nabla}\,\Theta(\vec{r})\,.$$

We easily recognize the superfluid velocity without magnetic field, $\vec{v}_s = (\hbar/m^*)\,\vec{\nabla}\,\Theta$, as given by (4.79) and we can write

$$\frac{\hbar}{i}\,\vec{\nabla}\,\psi(\vec{r}) = m^*\,|\psi|\,\vec{v}_s(\vec{r})\,e^{i\Theta(\vec{r})}\,. \tag{4.84}$$

Note that we have initially assumed that the particle velocity is constant in space, $|\vec{v}_s(\vec{r})| = v_s$. Thus, after inserting the above vector (4.84) into the Ginzburg-Landau free energy (4.83), the volume integral can simply be replaced by the volume V of the material. We obtain

$$F = F_{M,n} + V\left(\frac{m^*}{2}\,|\psi|^2\,v_s^2 + r\,|\psi|^2 + \frac{u}{2}\,|\psi|^4\right).$$

Here, one can nicely see that the term with the superfluid velocity describes the *kinetic energy* of the superconducting condensate.

Next, we continue with the minimization of $F(|\psi|, v_s)$ with respect to the parameter $|\psi|$. We find

$$\frac{\partial F}{\partial|\psi|} = V\big(m^* v_s^2\,|\psi| + 2r\,|\psi| + 2u\,|\psi|^3\big) = 0,$$

and it follows

$$m^* v_s^2 + 2r + 2u\,|\psi|^2 = 0.$$

Using $r < 0$ we obtain the following result for the superconducting order parameter,

$$|\psi| = \sqrt{\frac{|r| - \dfrac{m^* v_s^2}{2}}{u}}\,.$$

The behavior of $|\psi|$ as a function of the superfluid velocity v_s is illustrated in Figure 4.5. The order parameter value is reduced in the presence of a finite v_s and decreases as the value of v_s is increased. Furthermore, it is seen that at a certain critical velocity $v_c(T) = \sqrt{2|r(T)|/m^*}$ the superconducting state breaks down.

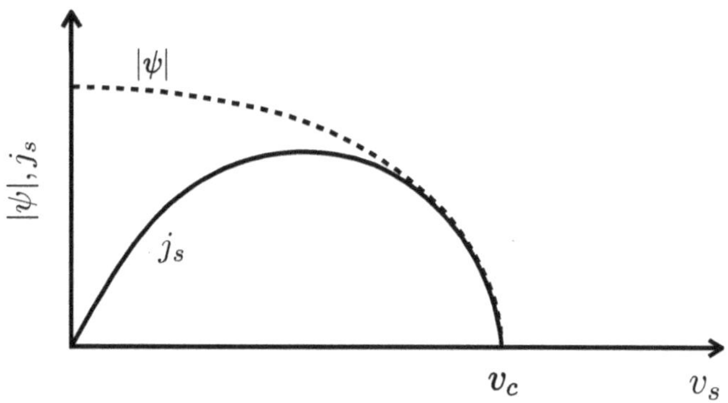

Figure 4.5: Superconducting order parameter $|\psi|$ (dashed line) and supercurrent j_s (solid line) as a function of the superfluid velocity v_s in a superconductor with a homogeneous supercurrent. The superconducting state breaks down at a certain critical velocity v_c. There is a finite velocity where the current becomes maximal.

In addition, it is interesting to look at the supercurrent $j_s = |\vec{j}_s|$ as a function of v_s. It is found from the expression

$$j_s = q^* |\psi|^2 v_s = q^* v_s \frac{|r| - \frac{m^* v_s^2}{2}}{u}, \tag{4.85}$$

where the resulting behavior is also illustrated in Figure 4.5. The appearance of a maximum supercurrent at a finite superfluid velocity originates from the proportionality of j_s not only to v_s but also to the density of Cooper pairs, $|\psi|^2$. The value of $|\psi|^2$ decreases more rapidly than v_s increases, which results in a maximum value of j_s with respect to variation of v_s.

4.5 – The characteristic length scales

Finally, let us discuss the characteristic length scales entering the Ginzburg-Landau theory via the free energy F.

4.5.1 – Inhomogeneous system, field-free case

At first, we consider again the field-free case $\vec{A} = 0$. According to (4.16), for small values of the order parameter ψ the most important contributions to the free energy are as follows,

$$F - F_n \propto \int_{\mathcal{V}} d^3r \left[\frac{\hbar^2}{2m^*} |\vec{\nabla}\psi(\vec{r})|^2 - |r||\psi(\vec{r})|^2 + \dots \right].$$

$$(4.86)$$

Better comparison of the different terms in (4.86) is achieved by factoring out $|r|$ such that all terms become dimensionless. We obtain

$$F - F_n \propto \int_{\mathcal{V}} d^3r \left[\frac{\hbar^2}{2m^*|r|} |\vec{\nabla}\psi(\vec{r})|^2 - |\psi(\vec{r})|^2 + \dots \right].$$

$$(4.87)$$

The length scale in which the order parameter changes is revealed by comparing the gradient term $\propto |\vec{\nabla}\psi|^2$ in (4.87) with the term $\propto |\psi|^2$. The length which we can read off from the pre-factor of the first term is

$$\xi = \frac{\hbar}{\sqrt{2m^*|r|}}.$$

$$(4.88)$$

This is exactly the Ginzburg-Landau coherence length found in section 4.2 where the Ginzburg-Landau equations of a system with translational symmetry in the presence of a surface were solved.

The behavior of ξ close to the transition temperature T_c is determined by $|r(T)| \propto (T_c - T)$ (compare section 4.1), and we find from (4.88) the proportionality

$$\xi(T) \propto (T_c - T)^{-1/2}.$$

$$(4.89)$$

This result is also known already from section 4.2.

Finally, to find the connection to the experiment, let us express ξ in terms of measurable quantities. Later we will show that between H_c and the parameters r and u there is the relationship

$$\frac{r^2}{u} = \frac{H_c^2}{4\pi}.$$

We can eliminate the parameter u with the help of the equation (4.6),

$$|r|/u = n_s^*,$$

and we obtain the expression

$$|r| = H_c^2/(4\pi n_s^*).$$

Inserting this equation into (4.88), we find for the Ginzburg-Landau coherence length

$$\xi = \sqrt{\frac{2\pi n_s^* \hbar^2}{m^* H_c^2}}, \tag{4.90}$$

which is the characteristic length in which the Cooper pair density changes in space. It can be determined experimentally by measuring the Cooper pair density in the bulk material and the critical magnetic field. In the following, we show that an *additional* length scale enters the theory if a magnetic field is applied.

4.5.2 – Homogeneous system in a magnetic field

To extract this length scale we look at the opposite of the previous case and fix the order parameter to a constant value, $\psi(\vec{r}) = \psi$, such that the coherence length ξ becomes unimportant here. In addition, we take into account a magnetic field, $\vec{A}(\vec{r}) \neq 0$. Note that this situation describes the case where the London theory is valid. Therefore, one can expect that the length scale sought here is the London penetration depth.

We start again from the free energy in the extended form (4.57),

$$F[\vec{A}, T, \psi] = F_{M,n} + \int_{\mathcal{V}} d^3r \left[-\frac{\hbar^2}{2m^*}\psi^* \vec{\nabla}^2 \psi \right.$$

$$-\frac{\hbar q^*}{ic\,2m^*}\left(\psi^*\vec{A}\cdot\vec{\nabla}\psi + \psi^*\vec{\nabla}\cdot(\psi\vec{A})\right)$$

$$+\frac{1}{2m^*}\left(\frac{q^*}{c}\right)^2 \vec{A}^2|\psi|^2 + r|\psi|^2$$

$$\left. +\frac{u}{2}|\psi|^4 + \frac{1}{8\pi}(\vec{\nabla}\times\vec{A})^2 \right],$$

and choose London's gauge (2.17) such that the vector potential fulfills the condition $\vec{\nabla}\cdot\vec{A} = 0$. Furthermore, our initial assumption of a constant ψ leads to $\vec{\nabla}\psi = 0$. Thus, the remaining contributions, which contain the vector potential and derivatives of it, are

$$F - F_{M,n} = \int_{\mathcal{V}} d^3r \left[\frac{q^{*2}}{2m^*c^2}n_s^*\vec{A}^2 + \frac{1}{8\pi}(\vec{\nabla}\times\vec{A})^2 \right] - \Delta E,$$

$$(4.91)$$

where $n_s^* = |\psi|^2$. The energy $\Delta E > 0$ with

$$\Delta E = V\left(|r||\psi|^2 - \frac{u}{2}|\psi|^4\right),$$

describes the energy gain of the homogeneous superconducting system (in the case where the magnetic field is switched off) compared to the same system in the normal state. It is responsible for the stability of the superconducting state.

We can compare the different terms in (4.91) containing $\vec{A}$ and its spatial derivatives again by excluding the preliminary factors. In (4.91), this corresponds to the factor in front of $\vec{A}^2$ in the first term. After excluding it we find

$$F - F_{M,n} = \frac{q^{*2}n_s^*}{2m^*c^2}\int_{\mathcal{V}} d^3r \left[\vec{A}^2 + \frac{m^*c^2}{4\pi q^{*2}n_s^*}(\vec{\nabla}\times\vec{A})^2 \right] - \Delta E.$$

Again, we read off the characteristic length scale from the pre-factor of the term containing spatial derivatives, i. e. the $(\vec{\nabla} \times \vec{A})^2$ term. The corresponding length is given by the square root of this factor,

$$\lambda_L = \sqrt{\frac{m^*c^2}{4\pi q^{*2}n_s^*}} \,. \tag{4.92}$$

The agreement with the London penetration depth (2.13) becomes clear after taking into account the relations $q^* = -2e$, $n_s = 2n_s^*$, and $m = m^*/2$ between Cooper pairs and electrons.

The temperature behavior of λ_L is determined by

$$n_s^*(T) = |\psi|^2 = |r|/u \propto (T_c - T)$$

(compare section 4.1). From (4.92) it follows

$$\lambda_L(T) \propto (T_c - T)^{-1/2}, \tag{4.93}$$

which means that the London penetration depth behaves near the critical point like ξ. Thus, for both λ_L and ξ the critical exponent is equal to (-1/2).

The above considerations show that there are two natural length scales within the Ginzburg-Landau theory, the coherence length ξ and the London penetration depth λ_L. Both lengths are intrinsic properties of the material, but both also depend on the temperature. However, the ratio of the two lengths, the Ginzburg-Landau parameter $\kappa = \lambda_L/\xi$, is *temperature-independent* due to (4.89) and (4.93). Therefore, this parameter can be considered as a real material quantity, serving to classify superconductors in type 1 and type 2 according to subsection 1.3.3.

Chapter 5 – Theoretical applications

In this chapter, important applications of the theories discussed so far are presented. Of particular interest is the behavior of the superconductor near the transition point in the presence of an external magnetic field. The most important aspect to be investigated is the phase transition from the superconducting state to the normal state, which is generated by a magnetic field. This means that the normal state is not established by increasing the temperature, but by an external magnetic field, whose field strength is systematically increased at a fixed temperature. Many experiments have shown that superconductors behave differently during this magnetic field-generated transition. While the conventional superconductors (metallic materials) show an abrupt transition to the normal state, the unconventional superconductors (in particular the high-temperature superconductors) form a typical intermediate state characterized by a partial penetration of the magnetic field into the material. With the help of the Ginzburg-Landau theory, it is possible to explain most of the phenomena occurring in this context. This should be the goal of the present chapter.

In the section 5.1, we first show a general method based on the Gibbs free energy, which can be used to calculate the critical field strength for the transition to the normal state. This method is then first applied to a simple model that we have already used often in the previous sections, the homogeneous superconductor. We calculate the critical field strength for this model in section 5.2.

In the next section 5.3 we explain some experimental observations for the phase transition in conventional superconductors. In this context, we study the latent heat and draw conclusions about the order of the phase transition.

The section 5.4 is dedicated to the investigation of a possible inhomogeneous superconducting state when an external magnetic field is applied. Here we investigate whether a partial penetration of the magnetic field into the superconductor can be thermodynamically stable. Using the method introduced in the section 5.1, it is shown that under a certain condition such a state is indeed possible. We will derive this condition and determine it precisely. It turns out, if this condition is met, that the transition to the normal state is then divided into two phase transitions with the respective critical field strengths H_{c1} and H_{c2}. In this context, the two known types of superconductors are defined. We will derive explicitly the value of the Ginzburg-Landau parameter κ which marks the boundary between these two types.

Based on the flux quantization phenomenon, we introduce in section 5.5 a model of Abrikosov vortices. This model is then used to calculate the value of the lower critical field H_{c1}. On the other hand, the transition at the upper critical point H_{c2} is described through an explicit solution of the linearized Ginzburg-Landau equations. In this context, we calculate the value of H_{c2}.

5.1 – Critical magnetic field

We start with a theoretical description of the phase transition from the superconducting state to the normal state when an external magnetic field is applied. Although the underlying microscopic mechanism of this transition remains unknown within the framework of our phenomenological theory, we can study the transition based on thermodynamic arguments.

The starting point is a spatial area in which there is a magnetic field $\vec{H}(\vec{r})$ generated by some external setup. We will call this field an 'external magnetic field' from now on. For simplification, we assume that this external magnetic field is homogeneous, i. e. $\vec{H} = H\vec{e}_z$ where H is the external parameter which describes the strength of the magnetic field. A material with the volume V, which can be either in the normal state or in the superconducting state, is introduced into this field, and we are interested in the reaction of the material to the external field. Of particular importance is to calculate the specific values of the magnetic field strength H at which normally conductive areas form in the superconductor or the material completely passes into the normal state.

5.1.1 – General approach

First, we explain how the desired phase transition is described within the framework of our theory and how a critical field strength can then be calculated.

We consider a situation where the material with a fixed temperature T is in a homogeneous magnetic field $\vec{H} = H\vec{e}_z$. We now ask ourselves under what condition the system passes by itself from a certain state '1' to another state '2' or vice versa while the state variables temperature and magnetic field are fixed to some values T and $\vec{H}$, respectively. The states '1' and '2' are each described by special fields $\psi_1(\vec{r})$ and $\psi_2(\vec{r})$ of the superconducting order parameter, which must be two possible solutions of the Ginzburg-Landau equations with given parameters T and $\vec{H}$. For example, ψ_1 could be the homogeneous solution (4.8) and ψ_2 the solution for the normal state, $\psi_2(\vec{r}) = 0$. This particular combination of states is studied in detail below.

According to the considerations of section 3.5, the thermodynamic potential with T and $\vec{H}$ as natural variables is the Gibbs free energy $G[T, \vec{H}]$. Because of (3.23), a change of G only results if either the temperature or the external magnetic field changes. Consequently, a phase transition in which the material passes from a phase '1' described by the state ψ_1 to another phase '2' described by the state ψ_2 is only possible if the Gibbs free energy of phase '1', $G_1[T, \vec{H}, \psi_1]$, is *equal* to the Gibbs free energy of phase '2', $G_2[T, \vec{H}, \psi_2]$. Note that both functions of the Gibbs free energy are taken at the same values of T and $\vec{H}$ and they differ from each other only in the states ψ_1 and ψ_2. This expresses exactly the thermodynamic equilibrium, because we had defined the equilibrium value of the order parameter precisely by the fact that with a small variation of the order parameter, the thermodynamic potential does not change.

Thus, the critical magnetic field H_c at a certain temperature T is defined as the particular value of the magnetic field for which both phases can coexist in thermodynamic equilibrium. This means that the Gibbs free energy taken at H_c and T must have the same value for the two different fields ψ_1 and ψ_2 of the order parameter,

$$G_1[T, \vec{H}_c, \psi_1] = G_2[T, \vec{H}_c, \psi_2] . \tag{5.1}$$

This equation can be used to calculate the critical field H_c. The easiest way to do this is to use the following method.

At first, one solves the Ginzburg-Landau equations for a given magnetic field H and temperature T resulting in the two solutions $\psi_1(\vec{r})$ and $\psi_2(\vec{r})$ of the order parameter. Note that, of course, more than two solutions are also possible. In this case, the procedure described below must be carried out in pairs for all solutions. As result one obtains the fields ψ_1 and ψ_2 both as a function of the parameters H and T.

In the next step, one calculates the Gibbs free energies using these solutions and one obtains the functions $G_1[T, H, \psi_1(H, T, \vec{r})]$ and $G_2[T, H, \psi_2(H, T, \vec{r})]$. Thus, at a fixed temperature they are functions of H. These functions can now be compared with each other. As long as G_1 and G_2 are different from each other, there is no phase transition and the system selects the phase with the smaller Gibbs free energy. At a certain critical point $H = H_c$ the two functions intersect and one can calculate from

$$G_1[T, H_c, \psi_1(H_c, T)] = G_2[T, H_c, \psi_2(H_c, T)]$$

the critical value H_c at a given T.

5.2 – The homogeneous state

We demonstrate the method using the example of the transition between a homogeneous superconducting state with a constant order parameter and the normal state. We have already found the homogeneous solution in section 4.1 where a magnetic field was not taken into account. In the presence of a magnetic field, this state can also be taken, when the London penetration depth of the material is small compared to its spacial extension, $\lambda_L \ll V^{1/3}$. In section

2.2 we have shown that in this case the magnetic induction inside the material is zero, $\vec{B} = 0$ (Meissner effect).

In the following, we demonstrate the method described above, starting from solving the Ginzburg-Landau equations (4.70)-(4.72) up to the calculation of the critical field strength. We start with the superconducting solution with constant order parameter, which we define as state '1'. Since both the external field and the order parameter are constant we can write $\vec{\nabla} \times \vec{H} = 0$ and $\vec{\nabla} \psi_1 = 0$. The gauge invariance can be used to make the order parameter real, $\psi_1^* = \psi_1$. Moreover, due to $\vec{B}_1 = 0$ the vector potential is also zero, $\vec{A}_1 = 0$. Thus, equation (4.72) is immediately fulfilled. Equations (4.70) and (4.71) lead to

$$0 = -\psi_1 + \frac{u}{|r|}|\psi_1|^2\psi_1,$$

and the solution is $\psi_1 = \sqrt{|r|/u}$. Note that this is the same result as in the case of the homogeneous solution (4.8) without magnetic field. With (4.9) and (4.10) we obtain the temperature dependence of the order parameter,

$$\psi_1(T) = \sqrt{\frac{\tilde{r}_0}{\tilde{u}_0}}\sqrt{T_c - T}.$$

Note that this particular solution does not depend on the external magnetic field H.

Next, we calculate the corresponding Gibbs free energy G_1 as a function of H and T. The Gibbs free energy is defined by equation (3.22). Since $\vec{B}_1 = 0$ we have $G_1 = F_1$. The free energy can be found from (4.47) and (4.48) with the help of a constant ψ_1,

$$G_1 = F_1 = F_{1,M} = F_{M,n} + V\left(r\psi_1^2 + \frac{u}{2}\psi_1^4\right),$$

and using $\psi_1 = \sqrt{|r|/u}$,

$$G_1 = F_{M,n} + V\left(r\,\frac{|r|}{u} + \frac{u}{2}\,\frac{|r|^2}{u^2} \right).$$

With $r < 0$ and $u > 0$ (compare section 4.1), we obtain at first the following result

$$G_1 = F_{M,n} - V\,\frac{|r|^2}{2u}. \tag{5.2}$$

The temperature dependence of the parameters is determined by the equations (4.9) and (4.10). With them the final result for $G_1(T, H, \psi_1(H, T))$ in the superconducting state reads

$$G_1(T, H, \psi_1(H, T)) = F_{M,n}(T) - V\,\frac{\tilde{r}_0^2}{2\tilde{u}_0}(T_c - T)^2. \tag{5.3}$$

The important finding is that the Gibbs free energy in this special state '1' is *independent* of the external magnetic field.

Next, we turn to the normal state which we define as state '2'. If the material cannot be magnetized in the normal state, which applies to most metals, the magnetic induction is equal to the external magnetic field, $\vec{B}_2 = \vec{H}$. In this case, the definition equation (3.22) for the Gibbs free energy provides

$$G_2 = F_2 - \frac{1}{4\pi}VH^2,$$

where $F_2 = F(T, B_2 = H, \psi_2 = 0)$. The free energy in the normal state can be taken again from (4.47) and (4.48) with $\psi_2 = 0$ and $\vec{B}_2 = \vec{H}$. We find

$$F_2 = \frac{1}{8\pi}VH^2 + F_{M,n}(T).$$

Thus, the Gibbs free energy in the normal state reads

$$G_2 = F_{M,n}(T) - \frac{1}{8\pi}VH^2. \tag{5.4}$$

In contrast to the superconducting state, the Gibbs free energy in the normal state does depend on the magnetic field.

If one compares equation (5.4) with (5.2) or (5.3), one can see that with very small fields (as well as with $H = 0$) the Gibbs free energy in the superconducting state is always *smaller* than in the normal state. Therefore, the superconducting state is taken in small fields.

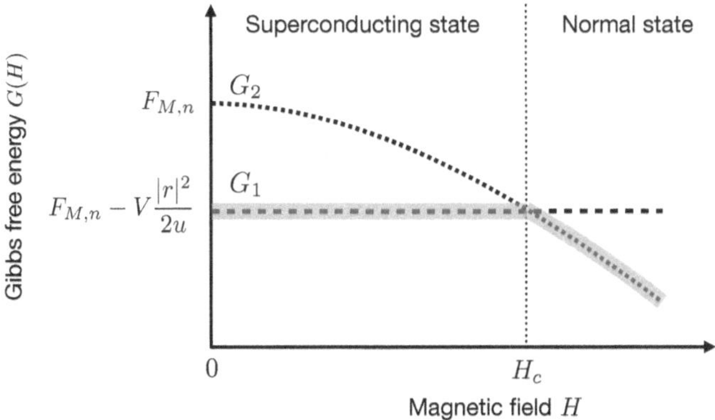

Figure 5.1: Gibbs free energies of the homogeneous superconducting state (G_1, dashed line) and the normal state (G_2, dotted line) as a function of the magnetic field H. At a given value of H the state with the lower value of the Gibbs free energy is 'chosen' by the system (highlighted with gray shade). Since the two curves intersect at a certain field strength H_c, the phase transition from the superconducting to the normal state takes place at H_c, which is manifested by a kink in the curve of Gibbs free energy.

The situation is shown in Figure 5.1 where the Gibbs free energies for both solutions are plotted separately as a function of H. While the Gibbs free energy G_1 for the superconducting state is independent of the magnetic field (dashed line), a parabolic decreasing course (dotted line) is found in the normal state according to equation (5.4). Both curves intersect at the critical point H_c, where the phase transition takes place. The state taken by the system is marked in gray in the figure. This shows that the Gibbs free energy at H_c has a kink, which is typical of a first-order phase transition.

The critical field H_c, where the superconducting state goes into the normal state, is defined by the equality $G_1(T, H_c) = G_2(T, H_c)$. It follows with the equations (5.4), (5.2) and (5.3)

$$F_{M,n} - V\frac{|r|^2}{2u} = F_{M,n} - \frac{1}{8\pi}VH_c^2 = F_{M,n} - V\frac{\tilde{r}_0^2}{2\tilde{u}_0}(T_c - T)^2 \,.$$

The solution reads

$$H_c(T) = \sqrt{\frac{4\pi r^2}{u}} = \sqrt{\frac{4\pi \tilde{r}_0^2}{\tilde{u}_0}}(T_c - T), \qquad (5.5)$$

which is valid for $T \le T_c$. Note that our theory is only valid for temperatures close to T_c from the outset. The critical field thus behaves linearly as a function of the temperature (close to T_c), which is shown in Figure 4.1 by the dashed line. In addition, this result can explain the right outer corner of the H-T phase diagram (see Figure 1.5), where the phase boundary runs in obliquely against T_c.

In summary, the following picture results for the phase transition. We consider a fixed temperature $T < T_c$ and slowly increase the magnetic field starting from $H = 0$. The system is initially in the superconducting state, with the order parameter taking the constant value

$$\psi_1 = \sqrt{|r|/u} = \sqrt{|\tilde{r}_0|/\tilde{u}_0}\sqrt{T_c - T}$$

regardless of H. The Gibbs free energy is also independent of H in the superconducting state and has the value (5.2) which can be also expressed in terms of the critical field using (5.5),

$$G_1 = F_{M,n} - V\frac{H_c^2}{8\pi} \,. \qquad (5.6)$$

Once the magnetic field has reached the critical value H_c, which is given by equation (5.5), the system suddenly drops into the normal state, i. e. the order parameter drops down to zero. From this point, the Gibbs free energy does depend on H and it follows the parabolic decreasing course, which is given by equation (5.4). With this, the Gibbs free energy shows a kink at the critical point. This behavior is typical of a first-order phase transition.

With the above condition (5.1), we have a determination equation at hand, which we can use to systematically calculate critical fields. To carry out this procedure in a systematic way, however, we first have to find the solutions of the Ginzburg-Landau equations and then, using these solutions, we have to solve (5.1) for H_c. Note that due to the non-linearity of the equation with respect to H_c, there can be several solutions that describe different types of phase transitions. A simplified approach to solving this problem will be examined in the section 5.4.

5.3 – Latent heat

Before we look at this approach, we should first discuss an important experimental observation, which can be well explained within our theory as a consequence of the transition into the homogeneous superconducting state.

A rather simple way to experimentally specify the type of the phase transition from the superconducting to the normal state is to investigate the *latent heat* which is exchanged during this transition. Measuring this quantity points out the nature (first order or second order) of the transition. While the latent heat for a second-order phase transition is absent, it is finite for a first order transition.

As already mentioned in the subsection 1.3.2, experiments have found that the transition to the superconducting phase is of *first order* for $T < T_c$. In this temperature range, experiments have measured an exchange of latent heat during the transition. We want to explain this phenomenon with the help of our theory by explicitly calculating the latent heat as a function of temperature.

The latent heat can be calculated from the entropy exchange during the transition. According to equation (3.24) the entropy $S(T, H)$ can be obtained by forming the partial derivative of G with respect to T (fixed H),

$$S(T,H) = -\left.\frac{\partial G}{\partial T}\right|_H .\tag{5.7}$$

For the study of the heat exchange during the transition it is necessary to consider the behavior of the entropy in the immediate vicinity of the phase boundary between the superconducting and normal state. The latent heat ΔQ can then easily be calculated from the difference of entropy values close to the phase boundary,

$$\Delta S = \lim_{\delta H \to 0} \left[S_1(T, H_c - \delta H) - S_2(T, H_c + \delta H) \right].\tag{5.8}$$

This formula describes the difference between a certain entropy value $S_1(T, H_c - \delta H)$ in the superconducting phase slightly below the critical field H_c and an entropy value $S_2(T, H_c + \delta H)$ in the normal phase slightly above the critical field. According to the second law of thermodynamics, $\Delta Q = T \Delta S$, a finite entropy difference $\Delta S \neq 0$ (discontinuity with respect to the field) is related to a finite latent heat ΔQ. In this case, the phase transition is of first order and accompanied by heat transfer with the environment of the material.

In the following, we show that the transition from the superconducting state '1' to the normal state '2' is of first order in the case $T < T_c$. Using (5.7) and (5.8), we find

$$\begin{aligned}
\Delta S = -\lim_{\delta H \to 0} &\left[\left.\frac{\partial G_1}{\partial T}\right|_{H=H_c(T)-\delta H} - \left.\frac{\partial G_2}{\partial T}\right|_{H=H_c(T)+\delta H} \right] \\
&= -\frac{\partial}{\partial T}(G_1 - G_2)\Big|_{H=H_c(T)},
\end{aligned}\tag{5.9}$$

where G_1 and G_2 are the Gibbs free energies in the superconducting and normal state, respectively. Both quantities are considered close to the phase boundary. Note that at the phase boundary ($\delta H = 0$) the relation $(G_1 - G_2)_{H=H_c(T)} = 0$ is found according to the above discussion. However, the difference of temperature derivatives may be non-zero, which leads to a discontinuity of the entropy. This can be seen well in Figure 5.1 at the different slopes of the red and blue curve at the point $H = H_c$.

To evaluate the temperature derivative in (5.9), we need the expression for $(G_1 - G_2)$ as a function of T, which can be derived from equations (5.3) and (5.4). We find

$$G_1 - G_2 = \frac{1}{8\pi}VH^2 - V\frac{\tilde{r}_0^2}{2\tilde{u}_0}(T_c - T)^2.$$

With the temperature derivative according to (5.9) we obtain for the change of entropy

$$\Delta S = -V\frac{\tilde{r}_0^2}{\tilde{u}_0}(T_c - T).$$

Finally, the relation $\Delta Q = T\Delta S$ results in the following expression for the latent heat,

$$\Delta Q = -V\frac{\tilde{r}_0^2}{\tilde{u}_0}T(T_c - T).$$

Consequently, the entropy change ΔS is finite for $T < T_c$ indicating a first order transition. On the other hand, the zero field transition at $T = T_c$ between superconducting and normal state is of second order.

5.4 – The inhomogeneous state

So far, we have described the superconducting phase in such a way that the density of the superconducting particles always remains constant, i. e. the state is homogeneous. By creating a magnetic field, this state does not change at first, but then abruptly and as a whole goes into the normal state at a critical field strength H_c. Such a behavior is evident for most metals, the so-called conventional superconductors.

However, there are materials in which a phase transition to a new superconducting state takes place even at a smaller field strength below H_c, which is fundamentally different from the homogeneous state discussed so far. As shown in the summary of the experiments in the subsection 1.3.3, this intermediate state has the character that superconducting and normally conductive regions occur alternately. So this state has an inhomogeneous character. Interestingly, this behavior can be found in most so-called high-temperature superconductors, which are important for technical applications.

In this section, we want to show that the Ginzburg-Landau theory is able to explain the behavior of such unconventional materials. To do this, we first find out that an inhomogeneous superconducting state can actually exist. Furthermore, we will show that such a state, in accordance with the experiments, only occurs for certain materials and that it is only thermodynamically stable in a certain field strength range.

We can describe different materials within the framework of the theory by varying the material parameters λ_L and ξ. In the following, we examine whether there are particular values (or regions of values) of these parameters for which an inhomogeneous solution $[\psi(\vec{r}), B(\vec{r})]$ of the Ginzburg-Landau equations (4.70)-(4.72) can be found, which occurs *in addition* to the usual solutions of the homogeneous superconducting state and the normal state (both discussed above). This is a promising approach since in the subsection 4.2.3 we already found such a state in the special case without magnetic field.

So we assume that the system can take up three different states as a function of the magnetic field H. We denote these states by the indices 1-3, respectively:

1. The homogeneous superconducting state $\psi_1 = \sqrt{|r|/u}$ and $B_1 = 0$.

2. The normal state $\psi_2 = 0$ and $B_2 = H$.

3. The inhomogeneous superconducting state with spatial dependent order parameter and magnetic induction, $\psi_3 = \psi_3(\vec{r})$ and $B_3 = B_3(\vec{r})$.

All of the cases (1-3) must each be solutions of the Ginzburg-Landau equations. We have already shown this for the cases (1) and (2) in the section 5.2.

We first want to refrain from explicitly solving the Ginzburg-Landau equations and instead first assume that an inhomogeneous solution '3' exists for all field strengths H (in addition to the solutions '1' and '2'). Under this assumption, we want to investigate in the following which of these three solutions is thermodynamically stable at which field strengths. The condition for which the solution '3' exists at all will be investigated later.

In order to decide which of the three states is taken by the system, it is necessary to compare the associated Gibbs free energies G_1, G_2, G_3 as a function of H. For the homogeneous superconducting state (1) and the normal state (2) the Gibbs free energies are given by (5.6) and (5.4), respectively. They read as a function of H,

$$G_1(H) = F_{M,n} - V\frac{H_c^2}{8\pi}, \quad G_2(H) = F_{M,n} - V\frac{H^2}{8\pi}. \tag{5.10}$$

The Gibbs free energy G_3 of the inhomogeneous superconducting state is obtained from the free energy taken at the superconducting solution $[\psi_3(\vec{r}), \vec{B}_3(\vec{r})]$ via the transformation (3.22),

$$G_3(H) = F_3[T, \vec{B}_3, \psi_3] - \frac{1}{4\pi}\int_{\mathcal{V}} d^3r\, \vec{B}_3(\vec{r}) \cdot \vec{H}. \tag{5.11}$$

The free energy F_3 in the inhomogeneous superconducting state is given by (4.47) and (4.48). We insert this expression in the above equation and obtain

$$G_3(H) = F_{M,n} + \int_{\mathscr{V}} d^3r \left[\frac{1}{2m^*} \left| \left(\frac{\hbar}{i} \vec{\nabla} - \frac{q^*}{c} \vec{A}_3 \right) \psi_3 \right|^2 \right.$$

$$\left. + r|\psi_3|^2 + \frac{u}{2}|\psi_3|^4 + \frac{1}{8\pi}\vec{B}_3^2 - \frac{1}{4\pi}\vec{B}_3 \cdot \vec{H} \right] . \tag{5.12}$$

In the following, we want to develop a qualitative picture of the phase transition into the inhomogeneous state '3'. To do this, we need to take a closer look at the course of the function $G_3(H)$. In order to simplify this discussion, we consider from now on a system with translational symmetry in two spatial directions. This means that all physical quantities (order parameter and magnetic field) depend only on one spatial variable, which we set as the x coordinate. This translational symmetry corresponds exactly to the conditions considered for the system in subsection 4.2.3 but here with an additional magnetic field directed parallel to the z-axis, $\vec{H} = H\vec{e}_z$. Thus, a possible inhomogeneous solution of the Ginzburg-Landau equations should also be only dependent on x, and it should have the form $[\psi_3(x), B_3(x)]$.

Let us discuss the qualitative behavior of such a solution with the help of the considerations from the section 4.5. There we found that both the order parameter and the magnetic induction can only vary within two typical length scales, the coherence length ξ and the London penetration depth λ_L. The result was that ξ is the characteristic length scale for the order parameter and λ_L is that for magnetic induction. For the translational symmetry considered here the two length scales both occur in x direction.

This results in the following qualitative behavior for $\psi_3(x)$ and $B_3(x)$, which is shown schematically in Figure 5.2. The figure shows the entire length of the material in x direction starting at $x = 0$ to $x = L$. Since the order parameter (black solid line) can only change within ξ, characteristic regions of extent ξ near the boundaries of the material are formed where the order parameter is strongly reduced compared to its bulk value ψ_1 (black dotted line). We have

118

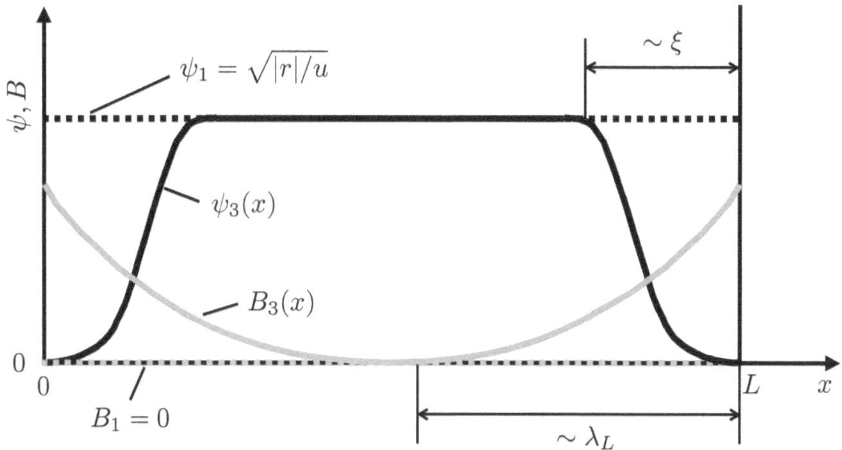

Figure 5.2: Sketch of a possible inhomogeneous superconducting solution of the order parameter $\psi_3(x)$ (black solid line) and the magnetic induction $B_3(x)$ (gray solid line) inside the superconducting material which expands in the x direction in the range $[0...L]$. The variables change within their characteristic length scales ξ and λ_L. The homogeneous superconducting solution $[\psi_1, B_1]$ is also shown (dotted lines).

already found this behavior of an inhomogeneous solution for the field-free case in Figure 4.3.

On the other hand, there is a typical penetration of magnetic induction into the material, which is described by the function $B_3(x)$. Starting from the edge of the material, the function falls exponentially within the length scale λ_L (compare Figure 2.1).

5.4.1 – Two phase transitions

With these findings, we can easily open up the desired behavior of the Gibbs free energy $G_3(H)$ in the inhomogeneous state. To do this, we look at equation (5.12). First of all, we see that the H dependence is dominated by the last term $\int d^3r(\vec{B}_3 \cdot \vec{H})/4\pi$, since it contains H explicitly. Since $B_3 \propto H$, G_3 falls parabolically similar to G_2 (Gibbs free energy for the normal state) with increasing H. However, the decrease is weaker than that of G_2, since $B_3 < H$ inside the material.

The parabolic fall of $G_3(H)$ is determined by the volume integral $\int d^3r(\vec{B}_3 \cdot \vec{H})/4\pi$. It becomes larger if the regions with a magnetic induction (compare the gray line in Figure 5.2) become broader. Since this width is determined by the London penetration depth λ_L, one can conclude that the parabolic fall of G_3 becomes stronger the larger λ_L is. This behavior is seen when the panels (a) and (b) of Figure 5.4 are compared.

Next, we discuss the $G_3(H)$ function at very small fields H. In particular, we are interested in the approached value for $H \rightarrow 0$. This value is determined by the terms of (5.12), which contain the absolute value $|\psi_3|$ of the order parameter. On the other hand, due to equation (5.11) or (3.22), in the limit $H \rightarrow 0$ the Gibbs free energy approaches the free energy, i. e. $G(H \rightarrow 0) = F$ applies.

That's why we will look at the free energy again next. The free energy of an inhomogeneous superconducting state was already calculated in subsection 4.2.5. In particular, with (4.42) we have evaluated the difference ΔF between the free energy in the inhomogeneous state '3' and the free energy in the homogeneous state '1'. Thus, in the notation used here, we get the connection

$$\Delta F = F_3 - F_1 = (G_3 - G_1)_{H \rightarrow 0}.$$

According to the calculated area-related value of ΔF, which is given by equation (4.44), we obtain the following proportionality,

$$(G_3 - G_1)_{H \rightarrow 0} \sim H_c^2 \, \xi.$$

It can be seen that the energetic distance between $G_3(H = 0)$ and $G_1(H = 0)$ is proportional to the coherence length ξ.

Furthermore, we obtain some important insights from equation (4.41) which describes the value of the free energy in the presence

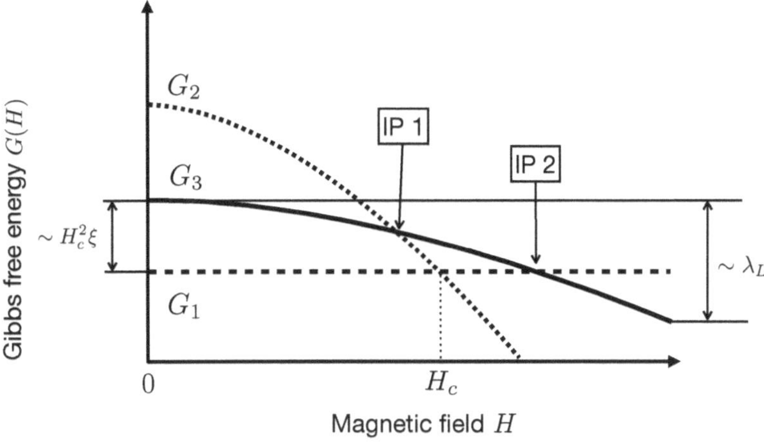

Figure 5.3: Schematic picture of the Gibbs free energy as a function of the magnetic field for the three different states. The function G_3 intersects the other two functions at two intersection points which are distributed left and right of the parameter H_c.

of a general superconducting solution with the order parameter $\psi(x)$: (i) Since $|\psi_3(x)| \geq 0$, we find for the relation to the normal state Gibbs free energy $G_3(H = 0) < G_2(H = 0)$. (ii) Since $|\psi_3(x)| \leq \psi_1$, the starting value of G_3 is larger than that of G_1, i. e. $G_3(H = 0) > G_1(H = 0)$.

In summary, the following qualitative picture is obtained for $G_3(H)$. It is illustrated in Figure 5.3 where $G_3(H)$ is shown by the black solid line. The function starts at $H = 0$ with a function value that is between the values $G_1(H = 0)$ and $G_2(H = 0)$. The difference between $G_3(H = 0)$ and $G_1(H = 0)$ (space between the thin lines on the left side of the plot) scales with $H_c^2 \xi$. With increasing magnetic field, $G_3(H)$ falls approximately parabolically. The decrease scales with λ_L (indicated by the space between the thin lines on the right side of the plot). However, this decrease is not as strong as that of $G_2(H)$ (dotted line), so these two functions intersect at a certain intersection point 'IP 1'. In addition, because of the location of the starting point above the G_1 level, there is a second intersection 'IP 2' with the constant function $G_1(H)$. As discussed above, such intersections can describe possible phase transitions depending on where they are positioned. This will be examined in the following.

Let us now have a closer look how the intersections IP 1 and IP 2 shift when the system parameters ξ and λ_L are varied. It makes sense to change the parameters in such a way that the critical field strength H_c remains fixed. This ensures that the intersection between the dotted and dashed line in Figure 5.3 always remains unchanged during the variation.

On the other hand, this also means that if H_c is fixed, the two parameters ξ and λ_L can no longer be set independently of each other. In this case, there must be a connection between these two parameters. In order to find this connection, we look again at the corresponding equations (4.90) and (4.92) for ξ and λ_L,

$$\xi = \sqrt{\frac{2\pi n_s^* \hbar^2}{m^* H_c^2}}, \qquad \lambda_L = \sqrt{\frac{m^* c^2}{4\pi q^{*2} n_s^*}}.$$

Multiplying these two equations with each other, we obtain the desired connection between ξ, λ_L, and H_c,

$$\xi \, \lambda_L = \frac{c \hbar}{\sqrt{2} q^* H_c}. \tag{5.13}$$

So if materials with equal H_c are compared with each other, the product of coherence length and London penetration depth must be constant for these materials. Thus, it makes sense in the following discussion to compare materials with large ξ and small λ_L with those materials where ξ is small and λ_L large, so that the product of the two parameters always remains the same.

5.4.2 – Types of superconductors

We start with a material with large ξ and small λ_L. This case is shown in Figure 5.4(a). The curve $G_3(H)$ (solid line) starts at a relatively large value and then runs only slightly down-sloping. This behavior leads to the fact that either $G_1(H)$ (dashed line) or $G_2(H)$ (dotted line) is always smaller than G_3. The respective state, which is taken by the system, is shaded in the figure in gray. In such a material, the inhomogeneous solution is never taken and the usual phase transition as described in the Figure 5.1, where the homogeneous superconducting state goes directly into the normal state at H_c, takes place. These superconductors are called type 1 superconductors.

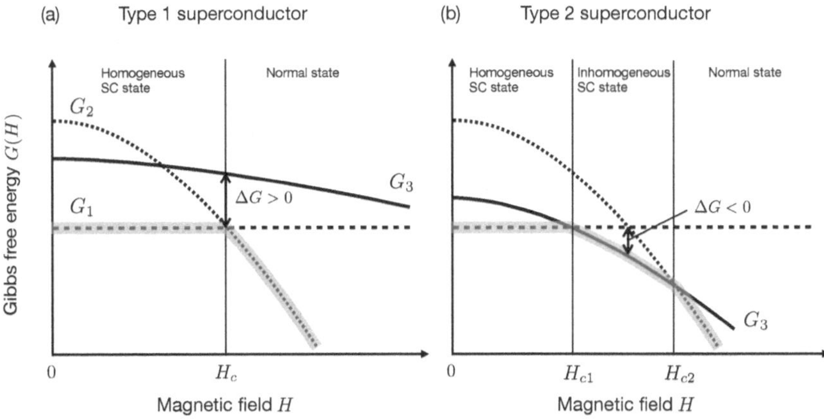

Figure 5.4: Schematic picture of the Gibbs free energy G_3 (solid line) for two different combinations of ξ and λ_L while the parameter H_c is kept fixed. (a) Large ξ and small λ_L (type 1 superconductor). The phase transition from the homogeneous superconducting state to the normal state takes place at H_c. The Gibbs free energy of the state taken by the system is shaded in gray. (b) Small ξ and large λ_L (type 2 superconductor). The two intersections of G_3 with G_1 and G_2 lead to two phase transitions at H_{c1} and H_{c2}.

In materials with relatively small ξ and relatively large λ_L, the image of the phase transition changes drastically. This case is shown in the Figure 5.4(b). Due to the small ξ, the $G_3(H)$ curve now starts at a relatively small value, which is still above the level of G_1. Thus, with very small field strengths, G_1 is still the smallest and the material therefore firstly takes the ordinary homogeneous superconducting state.

Due to the relatively large λ_L, the curve drops significantly more than in the case of panel (a), and as a result it can happen that at a certain critical value H_{c1}, which is smaller than H_c, the G_3 curve intersects the horizontal line G_1 even before the G_2 curve (normal state) intersects this line. As a result, starting at H_{c1} the value G_3 is the smallest, and thus the material shows a phase transition into an inhomogeneous superconducting state in which the magnetic field has partially penetrated the material.

If H is further increased, however, the G_3 curve intersects the G_2 curve at some further critical field value $H_{c2} > H_c$, because $G_2(H)$ falls faster than $G_3(H)$. From this point on, G_2 is the smallest and thus the system goes into the normal state at H_{c2}.

Thus, in materials with small ξ and large λ_L, there is a range between two critical field strengths H_{c1} and H_{c2}, in which an inhomogeneous superconducting state is taken, which is characterized by the fact that both the order parameter and the magnetic induction inside the superconductor are spatially dependent. Materials with this property are called type 2 superconductors.

Criterion for the type of superconductors

The considerations suggest that a good criterion for distinguishing type 1 and type 2 superconductors is the difference ΔG between the Gibbs free energy G_3 taken at H_c and G_1. This energy difference is marked in Figure 5.4 for both cases (a) and (b) each by a double arrow. It is defined by the following equation

$$\Delta G := G_3(H_c) - G_1. \tag{5.14}$$

Decisive for the type of superconductor is the *sign* of ΔG. So we distinguish the following two cases:

$\Delta G > 0$, type 1 superconductor

Either the homogeneous solution ψ_1 or the normal state have a lower Gibbs free energy than G_3. When the external field reaches the critical value H_c, a homogeneous transition from the superconducting to the normal conducting phase takes place in the way as discussed in section 5.2.

$\Delta G < 0$, type 2 superconductor

At H_c, the presence of an interface is energetically favorable and the system prefers an inhomogeneous superconducting state over a homogeneous one. Thus, at some critical value H_{c1}, which is smaller than H_c, a superconducting state in which both the order parameter and the magnetic induction vary spatially, as shown in Figure 5.2, becomes thermodynamically stable. The normal state is taken at a second critical field H_{c2} which is larger than H_c.

Next, let us explicitly evaluate the energy difference (5.14) and investigate under which condition ΔG is positive or negative. The Gibbs free energy $G_3(H_c)$ is obtained from equation (5.12) where $\vec{H}$ is replaced by $\vec{H}_c$,

$$
G_3(H_c) = F_{M,n} + \int_{\mathcal{V}} d^3r \left[\frac{1}{2m^*} \left| \left(\frac{\hbar}{i}\vec{\nabla} - \frac{q^*}{c}\vec{A}_3 \right) \psi_3 \right|^2 \right.
$$
$$
\left. + r|\psi_3|^2 + \frac{u}{2}|\psi_3|^4 + \frac{1}{8\pi}\vec{B}_3^2 - \frac{1}{4\pi}\vec{B}_3 \cdot \vec{H}_c \right], \qquad (5.15)
$$

and G_1 is given by the first equation of (5.10) where V is replaced again with the volume integral,

$$
G_1 = F_{M,n} - \frac{1}{8\pi}\int_{\mathcal{V}} d^3r \vec{H}_c^2 . \qquad (5.16)
$$

Note that the quantities $\vec{A}_3$, $\vec{B}_3$, and ψ_3 depend on the position vector $\vec{r}$. All these quantities refer to the inhomogeneous solution of the Ginzburg-Landau equations and therefore they all describe fields. In particular, the fields ψ_3 and $\vec{A}_3$ in (5.15) fulfill the Ginzburg-Landau equation (4.59),

$$\frac{1}{2m^*}\left(-i\hbar\vec{\nabla} - \frac{q^*}{c}\vec{A}_3\right)^2\psi_3 + r\psi_3 + u|\psi_3|^2\psi_3 = 0. \quad (5.17)$$

With the help of the equations (5.15) and (5.16) we can now find a general expression for the energy difference ΔG, which is defined by (5.14). We obtain

$$\Delta G = \int_{\mathcal{V}} d^3r\left[\frac{1}{2m^*}\left|\left(\frac{\hbar}{i}\vec{\nabla} - \frac{q^*}{c}\vec{A}_3\right)\psi_3\right|^2 + r|\psi_3|^2 + \frac{u}{2}|\psi_3|^4 \right.$$
$$\left. + \frac{\vec{B}_3^2}{8\pi} - \frac{1}{4\pi}\vec{B}_3 \cdot \vec{H}_c + \frac{\vec{H}_c^2}{8\pi}\right]$$

and after combining the last three terms,

$$\Delta G = \int_{\mathcal{V}} d^3r\left[\frac{1}{2m^*}\left|\left(\frac{\hbar}{i}\vec{\nabla} - \frac{q^*}{c}\vec{A}_3\right)\psi_3\right|^2 + r|\psi_3|^2 + \frac{u}{2}|\psi_3|^4 \right.$$
$$\left. + \frac{(\vec{B}_3 - \vec{H}_c)^2}{8\pi}\right]. \quad (5.18)$$

The first three terms can be simplified as well by use of the Ginzburg-Landau equation (5.17) as follows. Multiplying this equation by ψ_3^* and integrating over the volume $\mathcal{V}$, we find that the following volume integral vanishes,

$$\int_{\mathcal{V}} d^3r\left[\frac{1}{2m^*}\left|\left(-i\hbar\vec{\nabla} - \frac{q^*}{c}\vec{A}_3\right)\psi_3\right|^2 + r|\psi_3|^2 + u|\psi_3|^4\right] = 0.$$

Inserting this equation into (5.18), we obtain a simplified expression for ΔG,

$$\Delta G = \int_{\mathcal{V}} d^3r\left[\frac{(\vec{B}_3 - \vec{H}_c)^2}{8\pi} - \frac{u}{2}|\psi_3|^4\right].$$

It turns out to be useful to write the second term, which contains the order parameter, in terms of the homogeneous solution ψ_1,

$$\frac{u}{2}|\psi_3|^4 = \frac{u}{2}|\psi_1|^4 \frac{|\psi_3|^4}{|\psi_1|^4} = \frac{r^2}{2u}\frac{|\psi_3|^4}{|\psi_1|^4} = \frac{\vec{H}_c^2}{8\pi}\frac{|\psi_3|^4}{|\psi_1|^4},$$

where the known equations $|\psi_1|^2 = |r|/u$ and (5.5) were used. This leads us to the following expression,

$$\Delta G = \frac{1}{8\pi}\int_{\mathcal{V}} d^3r \left[(\vec{B}_3 - \vec{H}_c)^2 - \vec{H}_c^2\frac{|\psi_3|^4}{|\psi_1|^4}\right]. \tag{5.19}$$

Note that this expression is generally valid. It is not bound to translation symmetry.

In order to simplify the further calculation, however, from now on we will limit ourselves again to the translation symmetry introduced above. All quantities solely depend on x and the homogeneous magnetic field $\vec{H}$ points in z-direction. The material volume extends in the x direction in the range $[0\ldots L]$. Within this symmetry the volume integral in (5.19) reduces to an integral over x multiplied by the cross-sectional area A (taken in the y-z plane) of the material. With $\vec{B}_3 = B_3(x)\vec{e}_z$ and $\vec{H}_c = H_c\vec{e}_z$, the final result for the energy density $\Delta G/A$ in translational symmetry reads

$$\boxed{\frac{\Delta G}{A} = \frac{H_c^2}{8\pi}\int_0^L dx \left[\left(1 - \frac{B_3(x)}{H_c}\right)^2 - \frac{|\psi_3(x)|^4}{|\psi_1|^4}\right].} \tag{5.20}$$

The functions $B_3(x)$ and $\psi_3(x)$ in (5.20) together describe the inhomogeneous solution of the Ginzburg-Landau equations (4.70)-(4.72) for the case that the magnetic field is fixed to $H = H_c$. The associated functions for magnetic induction and order parameter have the form as shown schematically in Figure 5.2.

Since the case $H = H_c$ is considered in (5.20), $B_3(x)$ varies in the range $0 \le B_3(x) \le H_c$. On the other hand, $|\psi_3(x)|$ varies in the range $0 \leq |\psi_3(x)| \leq |\psi_1|$. Because of these properties, it is easy to see from (5.20) that ΔG can take both positive and negative values: The expression in square brackets, over which it is integrated, consists of a difference between two positive quadratic terms. The first one becomes large and the second one small (in this case $\Delta G > 0$) if both $B_3(x)$ and $|\psi_3(x)|$ remain small over the entire range in x direction. According to the Figure 5.2, this is the case when λ_L is rather small, but ξ rather large. This corresponds to the case of a type 1 superconductor shown in the Figure 5.4(a).

The reverse case $\Delta G < 0$ occurs according to equation (5.20) when $B_3(x)$ and $|\psi_3(x)|$ are increased over larger spatial areas. According to the Figure 5.2, this means that the magnetic induction penetrates the material in a relatively large area (gray solid line above the gray dotted line), while the regions with a reduced order parameter (black solid line below the black dotted line) become narrow. This, in turn, is the case with relatively large λ_L and relatively small ξ. One then gets exactly the case of a type 2 superconductor, which is shown in Figure 5.4(b).

The most important question to investigate is now: Under which condition for the parameters λ_L and ξ do we find a vanishing energy difference ΔG? According to the above discussion, this condition would then represent the boundary between type 1 ($\Delta G > 0$) and type 2 ($\Delta G < 0$) superconductors.

5.4.3 – Field equations in the London gauge

To answer this question, it makes sense to look for the particular solution of the Ginzburg-Landau equations that leads exactly to the equation $\Delta G = 0$. In order to achieve this goal, we start with a gauge transformation of these equations so that they are simplified, but their validity remains general. Then, in a second step, we try to find the special solution of these simplified equations which leads to $\Delta G = 0$.

As with most attempts to solve equations of motion, one first proceeds in such a way that one uses the existing freedoms to simplify the equations of motion as much as possible. In our case, this is the freedom to choose an appropriate local gauge transformation from which we know that it leaves the Ginzburg-Landau equations invariant (as we have shown in subsection 4.3.1). Therefore, let's take advantage of this freedom to simplify the problem by choosing the particular gauge in which the phase of the order parameter is identically zero, $\psi_3 = |\psi_3|$.

The formal process of the transformation is as follows. At first, we apply a special gauge transformation

$$\psi_3(\vec{r}) \rightarrow \psi_L(\vec{r}), \qquad \vec{A}_3(\vec{r}) \rightarrow \vec{A}_L(\vec{r}),$$

to the inhomogeneous solution, whereby the transformation is constructed in such a way that the new function $\psi_L(\vec{r})$ is real, i. e. it has the property $\psi_L(\vec{r}) = \psi_L^*(\vec{r})$. Such a transformation applied to the order parameter in the form $\psi_3(\vec{r}) = e^{i\Theta_3(\vec{r})}|\psi_3(\vec{r})|$ can be easily read off from the general form of the local gauge transformations, (4.50) and (4.51). We find,

$$\psi_L(\vec{r}) = e^{-i\Theta_3(\vec{r})}\psi_3(\vec{r}), \qquad \vec{A}_L(\vec{r}) = \vec{A}_3(\vec{r}) - \frac{\hbar c}{q^*}\vec{\nabla}\Theta_3(\vec{r}).$$

We have shown in subsection 4.3.1 that under a general local gauge transformation the free energy remains unchanged. Therefore, we can claim that the equation $F[\vec{A}_3, T, \psi_3] = F[\vec{A}_L, T, \psi_L]$ applies to the current transformation. Consequently, the Ginzburg-Landau equations retain their form (4.70)-(4.72), even with the new fields $\vec{A}_L$ and ψ_L. Note that both fields are now real, so that a simplification of the equations (4.70)-(4.72) will result.

We first look at the last of the three equations, the equation (4.72). First, since we have a homogeneous external magnetic field, $\vec{\nabla} \times \vec{H} = 0$ applies. Second, because of $\psi_L^* = \psi_L$, the first two terms on the right hand side of (4.72) cancel each other. These two properties taken together simplify the equation (4.72) as follows

$$\frac{q^{*2}}{m^*c^2}\psi_L^2\vec{A}_L + \frac{1}{4\pi}\vec{\nabla} \times (\vec{\nabla} \times \vec{A}_L) = 0. \tag{5.21}$$

Next, forming the curl of (5.21) and applying the known rules for the Nabla operator, we can reintroduce the magnetic induction

$$\vec{B} = \vec{\nabla} \times \vec{A}_L.$$

We find

$$
\begin{aligned}
0 &= \frac{q^{*2}}{m^*c^2}\vec{\nabla} \times (\psi_L^2 \vec{A}_L) + \frac{1}{4\pi}\vec{\nabla} \times (\vec{\nabla} \times \vec{B}) \\
&= \frac{q^{*2}}{m^*c^2}\left[(\vec{\nabla}\psi_L^2) \times \vec{A}_L + \psi_L^2 \vec{B}\right] \\
&\quad + \frac{1}{4\pi}\left[\vec{\nabla}(\vec{\nabla} \cdot \vec{B}) - \vec{\nabla}^2 \vec{B}\right].
\end{aligned}
\tag{5.22}
$$

Equation (5.21) provides an explicit formula for the vector potential,

$$\vec{A}_L = -\frac{m^*c^2}{4\pi q^{*2}\psi_L^2}\vec{\nabla} \times \vec{B},
\tag{5.23}$$

which we now use to eliminate $\vec{A}_L$ in (5.22). With the help of $\vec{\nabla} \cdot \vec{B} = 0$, we find

$$-\frac{m^*c^2}{4\pi q^{*2}\psi_L^2}(\vec{\nabla}\psi_L^2) \times (\vec{\nabla} \times \vec{B}) + \psi_L^2 \vec{B} = \frac{m^*c^2}{4\pi q^{*2}}\vec{\nabla}^2\vec{B}.
\tag{5.24}$$

At this point, it makes sense again to relate the order parameter and the magnetic induction to the fixed parameters $|\psi_1|$ and H_c in order to continue with the dimensionless variables

$$f(\vec{r}) = \frac{\psi_L(\vec{r})}{|\psi_1|}$$

and

$$\vec{b}(\vec{r}) = \frac{\vec{B}(\vec{r})}{H_c}.$$

We can expect from this action that a single length parameter can be extracted again, which describes the characteristic length scale in which the relevant quantities change spatially. We can easily introduce the dimensionless variables by dividing the equation (5.24) on both sides by $(H_c|\psi_1|^2)$. We first get

$$-\frac{m^*c^2}{4\pi q^{*2}\psi_L^2}(\vec{\nabla}f^2)\times(\vec{\nabla}\times\vec{b})+f^2\vec{b} = \frac{m^*c^2}{4\pi q^{*2}|\psi_1|^2}\vec{\nabla}^2\vec{b}\,.$$

In the first term, we still have to express ψ_L^2 by f. This is achieved by extending the fraction with $|\psi_1|^2$. We find

$$-\frac{m^*c^2}{4\pi q^{*2}|\psi_1|^2}\left(\frac{\vec{\nabla}f^2}{f^2}\right)\times(\vec{\nabla}\times\vec{b})+f^2\vec{b} = \frac{m^*c^2}{4\pi q^{*2}|\psi_1|^2}\vec{\nabla}^2\vec{b}\,.$$

We see that a spatial parameter is extracted as expected. It appears as a pre-factor in the terms that contain spatial derivatives and is given by

$$\lambda_L = \sqrt{\frac{m^*c^2}{4\pi q^{*2}|\psi_1|^2}}\,. \tag{5.25}$$

This parameter can be identified again as the London penetration depth (compare equation (4.92) with $n_s^* = |\psi_1|^2 = |r|/u$). Thus, with the property $\vec{\nabla}f^2 = 2f\,\vec{\nabla}f$ we arrive at the following differential equation for the real functions $f(\vec{r})$ and $\vec{b}(\vec{r})$,

$$\lambda_L^2\vec{\nabla}^2\vec{b} = -2\lambda_L^2\frac{\vec{\nabla}f}{f}\times(\vec{\nabla}\times\vec{b})+f^2\vec{b}\,. \tag{5.26}$$

This is the first of the Ginzburg-Landau equations in the particular gauge leading to a real order parameter function. Note that this equation is universal. It has only received a special shape as a result of the gauge transformation.

The second of the Ginzburg-Landau equations in this gauge is found by writing (4.70) or (4.71) (both equations are now equivalent since $\psi_L^* = \psi_L$) in terms of the transformed fields,

$$-\xi^2\left(\vec{\nabla}-\frac{iq^*}{\hbar c}\vec{A}_L\right)\cdot\left(\vec{\nabla}-\frac{iq^*}{\hbar c}\vec{A}_L\right)\psi_L - \psi_L + \frac{u}{|r|}\psi_L^3 = 0,$$

The execution of the differential operator with the help of the product rule leads to an extended form,

$$0 = -\xi^2\left[\vec{\nabla}^2\psi_L - \frac{q^{*2}}{\hbar^2 c^2}\vec{A}_L^2\psi_L - \frac{iq^*}{\hbar c}\left(\vec{\nabla}\cdot(\vec{A}_L\psi_L) + \vec{A}_L\cdot\vec{\nabla}\psi_L\right)\right] - \psi_L + \frac{u}{|r|}\psi_L^3. \quad (5.27)$$

We want to further simplify this equation by showing that the term in the round bracket vanishes,

$$\vec{\nabla}\cdot(\vec{A}_L\psi_L) + \vec{A}_L\cdot\vec{\nabla}\psi_L = 0. \quad (5.28)$$

To prove this, we take the first term $\vec{\nabla}\cdot(\vec{A}_L\psi_L)$ and replace the vector $\vec{A}_L$ with the help of equation (5.23). After that, the product rule for the Nabla operator is applied. We obtain

$$\begin{aligned}
\vec{\nabla}\cdot(\vec{A}_L\psi_L) &= -\frac{m^*c^2}{4\pi q^{*2}}\vec{\nabla}\cdot\left(\frac{1}{\psi_L}\vec{\nabla}\times\vec{B}\right)\\
&= \frac{m^*c^2}{4\pi q^{*2}}\left[\frac{1}{\psi_L^2}(\vec{\nabla}\times\vec{B})\cdot\vec{\nabla}\psi_L - \frac{1}{\psi_L}\vec{\nabla}\cdot(\vec{\nabla}\times\vec{B})\right].
\end{aligned}$$

The second term vanishes since $\nabla\cdot(\vec{\nabla}\times\vec{B}) = 0$ holds for any vector field. If we now replace in the first term the expression $\vec{\nabla}\times\vec{B}$ with the vector $\vec{A}_L$ using again (5.23), we immediately get $\vec{\nabla}\cdot(\vec{A}_L\psi_L) = -\vec{A}_L\cdot\vec{\nabla}\psi_L$. This proves the validity of the equation (5.28). With that, the Ginzburg-Landau equation (5.27) simplifies to

$$0 = -\xi^2\vec{\nabla}^2\psi_L + \frac{q^{*2}\xi^2}{\hbar^2 c^2}\vec{A}_L^2\psi_L - \psi_L + \frac{u}{|r|}\psi_L^3.$$

For the evaluation of this differential equation, it is advantageous to replace the vector potential in the second term with the magnetic induction which can be done by using equation (5.23) again. We find

$$0 = -\xi^2 \vec{\nabla}^2 \psi_L + \frac{q^{*2}\xi^2}{\hbar^2 c^2}\left(\frac{m^*c^2}{4\pi q^{*2}\psi_L^2}\right)^2 \psi_L(\vec{\nabla}\times\vec{B})^2$$

$$- \psi_L + \frac{u}{|r|}\psi_L^3$$

$$= -\xi^2 \vec{\nabla}^2 \psi_L + \left(\frac{\xi m^* c}{4\pi q^*\hbar}\right)^2 \frac{(\vec{\nabla}\times\vec{B})^2}{\psi_L^3} - \psi_L + \frac{u}{|r|}\psi_L^3.$$

In the next step, we again introduce the dimensionless variables $f = \psi_L/|\psi_1|$ and $\vec{b} = \vec{B}/H_c$. This should again create a common pre-factor, which then acts as a relevant length scale. The easiest way to achieve this is to divide the equation by $|\psi_1|$ and to extend the second term by H_c^2. This leads us to

$$0 = -\xi^2 \vec{\nabla}^2 f + \left(\frac{\xi m^* c H_c}{4\pi q^*\hbar}\right)^2 \frac{(\vec{\nabla}\times\vec{b})^2}{\psi_L^3 |\psi_1|} - f + \frac{u}{|r|}\psi_L^2 f. \tag{5.29}$$

The remaining old variable ψ_L in the second and fourth term can be replaced with f by means of $\psi_L = f|\psi_1|$ which leads us to

$$0 = -\xi^2 \vec{\nabla}^2 f + \left(\frac{\xi m^* c H_c}{4\pi q^*\hbar|\psi_1|^2}\right)^2 \frac{(\vec{\nabla}\times\vec{b})^2}{f^3}$$

$$- f + \frac{u}{|r|}|\psi_1|^2 f^3.$$

We can achieve a further simplification of this equation by eliminating the parameter $|\psi_1|^2$ in the second and fourth terms. In the second term, we replace $|\psi_1|^2$ with the equation

$$|\psi_1|^2 = \frac{\xi^2 m^* H_c^2}{2\pi\hbar^2},$$

which is found from equation (4.90) using $n_s^* = |\psi_1|^2$. In the fourth term, the best simplification is obtained by eliminating $|\psi_1|^2$ with the help of the known relation $|\psi_1|^2 = |r|/u$. So using these two replacements, we get

$$0 = -\xi^2 \vec{\nabla}^2 f + \left(\frac{c\hbar}{2q^*\xi H_c}\right)^2 \frac{(\vec{\nabla} \times \vec{b})^2}{f^3} - f + f^3.$$

The second term is further simplified with the help of equation (5.13) which can be written in the form

$$\frac{\lambda_L}{\sqrt{2}} = \frac{c\hbar}{2q^*\xi H_c}.$$

Finally, this leads us to the second Ginzburg-Landau equation in the reduced variables f and $\vec{b}$,

$$-\xi^2 \vec{\nabla}^2 f - f + f^3 + \lambda_L^2 \frac{(\vec{\nabla} \times \vec{b})^2}{2f^3} = 0. \tag{5.30}$$

The two coupled differential equations (5.26) and (5.30) are the Ginzburg-Landau equations in the particular gauge where the superconducting order parameter (represented by the function f) is real. Due to their general validity, they can be regarded as a system of basic equations for the superconducting state. Our goal is to use these equations to evaluate the expression (5.20) to find a condition for the parameters ξ and λ_L that leads to the equation $\Delta G = 0$. Since the expression (5.20) is limited to a system with translational symmetry, we must next formulate the differential equations (5.26) and (5.30) also in this symmetry.

5.4.4 – Boundary between type 1 and type 2

The translational symmetry considered here means that all variables depend only on x. In addition, the vector $\vec{H}$ of the homogeneous external magnetic field points in the z direction, which is to be written as $\vec{H} = H\vec{e}_z$. For the introduced dimensionless variables, this means

$$f(\vec{r}) = f(x)$$

and

$$\vec{b}(\vec{r}) = b(x)\vec{e}_z.$$

Thus, the Laplace operator in equation (5.30) can be written in the form $\vec{\nabla}^2 = \partial^2/\partial x^2$, and the curl of $\vec{b}$ reduces to $\vec{\nabla} \times \vec{b} = -b'(x)\vec{e}_y$. With these replacements equation (5.30) reads in the translational symmetry

$$-\xi^2 f'' - f + f^3 + \frac{\lambda_L^2}{2f^3}(b')^2 = 0. \tag{5.31}$$

For the second differential equation (5.26), the following vector equation results after inserting the above relations in translational symmetry,

$$\lambda_L^2 b'' \vec{e}_z + 2\lambda_L^2 \frac{f' \vec{e}_x}{f} \times (-b'\vec{e}_y) - f^2 b_z \vec{e}_z = 0.$$

This equation contains only a single equation for the z component. It reads

$$\lambda_L^2 b'' - 2\lambda_L^2 \frac{f'}{f} b' - f^2 b = 0. \tag{5.32}$$

From (5.31) and (5.32) we see again that the relevant parameters for the superconducting state are the Ginzburg-Landau coherence length ξ and the London penetration depth λ_L.

We now want to use these differential equations for the evaluation of the expression (5.20) for ΔG. The basic idea is to check which condition the parameters ξ and λ_L must meet so that the dimensionless variables $B_3(x)/H_c$ and $|\psi_3(x)|/|\psi_1|$ contained in the integral of (5.20) are solutions of the above differential equations and at the same time also result in $\Delta G = 0$. The easiest way to implement this idea is the following procedure. We first formulate a suitable ansatz that makes ΔG disappear. In a second step, we check under what condition this ansatz meets the equations (5.31) and (5.32).

We start with the first step, the formulation of a suitable ansatz: The right side of (5.20) would disappear if the function over which it is integrated were zero for all x, i. e. if

$$\left(1-\frac{B_3(x)}{H_c}\right)^2-\frac{|\psi_3(x)|^4}{|\psi_1|^4}=0$$

would be valid for all x, then one would find $\Delta G = 0$. Thus, a suitable ansatz for the relationship between $B_3(x)$ and $\psi_3(x)$ would be:

$$B_3(x) = \left(1-\frac{|\psi_3(x)|^2}{|\psi_1|^2}\right)H_c. \tag{5.33}$$

The second step is to determine the particular parameter values ξ and λ_L for which the ansatz (5.33) is a solution of equations (5.31) and (5.32). Written in the reduced variables

$$b = B_3/H_c$$

and

$$f = |\psi_3|/|\psi_1|,$$

equation (5.33) takes the simple form $b(x) = 1 - f^2(x)$. We use this relation to replace the variable b and all its derivatives with the variable f. For the first Ginzburg-Landau equation (5.31), we obtain with $b' = -2ff'$ the following equation

$$-\xi^2 f'' -f +f^3 +\frac{\lambda_L^2}{2f^3}(-2ff')^2 = -\xi^2 f'' -f +f^3 +2\lambda_L^2\frac{f'^2}{f} = 0.$$

Multiplication of this equation by f gives

$$0 = -\xi^2 f'' f -f^2 +f^4 +2\lambda_L^2 f'^2. \tag{5.34}$$

The second Ginzburg-Landau equation (5.32) takes a similar form after using the replacement $b'' = -2(f'^2 +ff'')$,

$$
\begin{aligned}
0 &= \lambda_L^2 b'' - 2\lambda_L^2 \frac{f'}{f} b' - f^2 b \\
&= -2\lambda_L^2(f'^2 + ff'') - 2\lambda_L^2 \frac{f'}{f}(-2ff') - f^2(1 - f^2) \\
&= -2\lambda_L^2 f'' f - f^2 + f^4 + 2\lambda_L^2 f'^2 .
\end{aligned}
\tag{5.35}
$$

Comparing the equations (5.34) and (5.35), we conclude that the ansatz (5.33) fulfills both of the Ginzburg-Landau equations only if $\xi^2 = 2\lambda_L^2$, or

$$
\frac{\lambda_L}{\xi} = \frac{1}{\sqrt{2}} .
\tag{5.36}
$$

This result is the searched condition that the parameters λ_L and ξ must meet in order for the equation $\Delta G = 0$ to be met. Therefore, this relationship has a special significance for experiments on superconductors, as it describes the 'boundary' between type 1 and type 2 superconductors.

5.4.5 – Inhomogeneous solution

For this particular case, in which the parameter values meet the equation (5.36), we now want to solve the Ginzburg-Landau equations approximately analytically. The aim is to find an inhomogeneous solution for the superconducting state in which the magnetic field partially penetrates the material. We will recognize that many of the qualitative considerations already developed in the previous subsections are confirmed.

If a special material whose parameters λ_L and ξ meet the equation (5.36) is in an external magnetic field with the field strength $H = H_c$, the corresponding inhomogeneous superconducting solution $[\psi_3(x), B_3(x)]$ meets for all x the sum rule

$$
\frac{B_3(x)}{H_c} + \frac{|\psi_3(x)|^2}{|\psi_1|^2} = 1,
\tag{5.37}
$$

and the renormalized order parameter $f(x) = \psi_3(x)/|\psi_1|$ is determined by the non-linear differential equation

$$0 = \xi^2(f'^2 - f''f) - f^2 + f^4.\tag{5.38}$$

The complete solution of this differential equation can only be found numerically. Nevertheless, the following findings can be gained from qualitative considerations. At first, it is easy to see that if f is constant ($f' = 0$), immediately $f = 1$ applies. The sum rule then leads to $B_3 = 0$. That is, if ψ_3 is spatially constant in any area of space, then $|\psi_3| = |\psi_1|$ (bulk value of the homogeneous solution) applies and the magnetic induction is zero in this space area.

As soon as the magnetic field penetrates into the material at a certain x ($B_3(x) > 0$), then the sum rule tells us that $|\psi_3(x)| < |\psi_1|$, i. e. the order parameter is reduced at this x compared to its homogeneous value $|\psi_1|$. Furthermore, one can see from the differential equation (5.38) that if $\psi_3(x)$ changes spatially ($f' \neq 0$), this change takes place on the length scale ξ.

The differential equation for the renormalized magnetic induction b can be found from (5.32) where we replace f with b again with the help of the sum rule $b + f^2 = 1$. From that one finds $f'/f = -b'/(2f^2)$ which is used in equation (5.32) to eliminate the terms with f and f'. We obtain at first

$$\lambda_L^2(1 - b)b'' + \lambda_L^2 b'^2 - (1 - b)^2 b = 0.$$

This equation is further simplified by dividing both sides by $(1 - b)^2$ and introducing a differential operator. Then we get the following differential equation for b,

$$\lambda_L^2 \frac{\mathrm{d}}{\mathrm{d}x} \frac{b'}{1 - b} - b = 0.\tag{5.39}$$

Equation (5.38) can be simplified in the same way. We obtain

$$\xi^2 \frac{\mathrm{d}}{\mathrm{d}x} \frac{f'}{f} + 1 - f^2 = 0.\tag{5.40}$$

The two differential equations (5.39) and (5.40) fully describe the inhomogeneous superconducting state for the boundary case

$$\xi = \sqrt{2}\lambda_L.$$

As already mentioned, the exact solution of (5.39) or (5.40) can be found numerically. Let us discuss here only an approximate solution for small fields $b \ll 1$, which should be valid inside the material. In this limiting case, (5.39) becomes

$$b'' - \frac{1}{\lambda_L^2}b = 0.$$

In order to be able to specify a solution, we still have to set boundary conditions. We define that b inside the material is constant at $x = L/2$ and, in accordance with our prerequisite, has a very small value there, $b(L/2) = b_0 \ll 1$. With these conditions, the solution reads

$$b(x) = \frac{b_0}{2}\left(e^{\frac{x-L/2}{\lambda_L}} + e^{-\frac{x-L/2}{\lambda_L}}\right) = b_0 \cosh\left(\frac{x-L/2}{\lambda_L}\right).$$

Note that strictly speaking, this solution is only valid in the region where $b(x) \ll 1$. The course of the cosh-function makes it easy to see that the qualitative image of Figure 5.2 is confirmed: The function $b(x)$ has a minimum in the center of the material and rises towards the edges. This increase occurs on the length scale of the London penetration depth λ_L. The solution for the order parameter function $f(x)$ can be found immediately from the sum rule $b + f^2 = 1$. One can see that the order parameter has a maximum in the center and falls towards the edges. The change also takes place within the length scale λ_L, which is also of the order of the coherence length ξ because of the selected relation $\xi = \sqrt{2}\lambda_L$.

5.4.6 – The Ginzburg-Landau parameter

In summary, it can be stated that the ratio λ_L/ξ is decisive for the basic properties of a superconductor in the magnetic field. Therefo-

re, this relationship is given its own name Ginzburg-Landau parameter,

$$\kappa := \frac{\lambda_L}{\xi}.$$

The parameter serves to distinguish the two types of superconductors. They are classified by their ability to form an inhomogeneous superconducting state which is characterized by the presence of regions in the bulk of the material where the magnetic field is present in form of a penetrated magnetic induction and where the superconducting order parameter is strongly reduced.

Superconductors of type 1 with $\kappa < 1/\sqrt{2}$ are characterized by a difference of the Gibbs free energy, $\Delta G > 0$, which means that regions where the magnetic field penetrates the material are energetically unfavorable. In these materials, the phase transition takes place abruptly at H_c and the homogeneous superconducting state passes directly into the normal state. For example, materials with vanishing λ_L values, as treated within the London theory in chapter 2, are prototypes of type 1 superconductors. It should also be noted that most conventional superconductors are of this type.

On the other hand, type 2 superconductors with the property $\kappa > 1/\sqrt{2}$ are characterized by negative energy difference, $\Delta G < 0$, such that a spatial variation of magnetic induction and order parameter becomes favorable. The phase transition into such an inhomogeneous state already takes place at a critical field strength H_{c1}, which is smaller than the parameter H_c. On the other hand, the transition to the normal state only takes place at a second critical value H_{c2}, which is larger than H_c. It is important to note that almost all high-temperature superconductors are type 2 superconductors. Therefore, we want to deal more intensively with this type in the next section.

5.5 – Model of type 2 superconductors

In the previous section, we have shown that the type 2 superconductors are particularly interesting because they show two phase transitions, between which an interesting state develops, in which the superconducting order parameter has a spatial structure. This behavior is well in line with the experiments. It has been shown that the famous high-temperature superconductors in particular are of this type. However, so far we have only been able to describe the phase transition at a qualitative level. Therefore, in this section we want to develop a model for the type 2 superconductors that correctly describes the inhomogeneous state in a wide range of the field strength. With this model, we then want to quantitatively determine the critical values H_{c1} and H_{c2}.

The inhomogeneous state is shown schematically for a material with translational symmetry in Figure 5.2. The tendency to form this state occurs because it is energetically favorable for the system to let the magnetic field partially penetrate near an interface to a non-superconducting region. This is shown in panel (b) of Figure 5.4, where one can see that in a certain field range, the Gibbs free energy G_3 belonging to this state takes the smallest value among all other possible states.

In the example of the qualitative illustration of this state in Figure 5.2, the extent L in one spatial direction is comparable to the London penetration depth λ_L. However, if L is much larger than the length scales λ_L and ξ, which is usually the case, the material tends to allow the magnetic field not only to penetrate at the edges, but also areas inside (far away from the edges) are formed in which a magnetic induction is present on a length scale λ_L. In these regions, which are called magnetic domains, the order parameter is then strongly reduced and the material can locally form regions where the normal state is present. This behavior should by described by the Ginzburg-Landau equations for a general three-dimensional system.

The aim of this section is to develop a model for the inhomogeneous state in type 2 superconductors which avoids a complete solution of the Ginzburg-Landau equations. With the help of this model, we calculate the critical field strengths H_{c1} and H_{c2} and de-

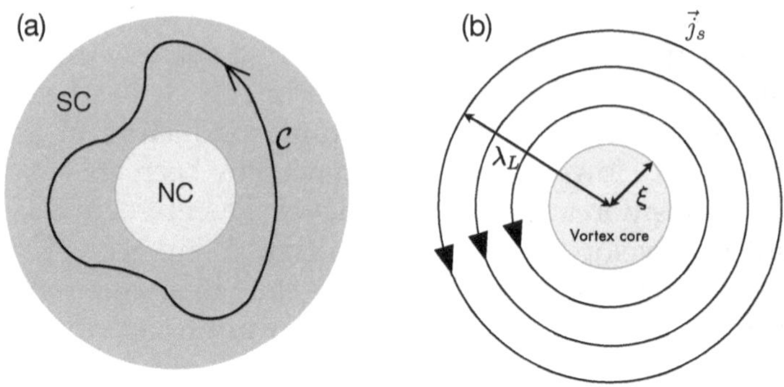

Figure 5.5:(a) Quantized magnetic flux through a normal conducting (NC) region which is enclosed by an arbitrary loop $\mathcal{C}$ inside the superconducting (SC) region. (b) Model of a vortex in type 2 superconductors. It consists of a cylindrical normal conducting vortex core (gray area with radius ξ) centered around the z axis. Circular supercurrents decay within the length λ_L.

scribe the structure of the inhomogeneous state for a two-dimensional system.

5.5.1 – Flux quantization

We start with the investigation of the inhomogeneous state near the lower critical field strength H_{c1}. Instead of solving the Ginzburg-Landau equations for a general situation, we want to pursue another much easier approach here, which is based on a model of isolated normal-conducting domains in a superconducting environment.

Since at H_{c1} the transition takes place starting from the pure homogeneous superconducting state (notation state '1'), it makes sense to also use such an environment as a starting point for the inhomogeneous state '3' near H_{c1}. So we first consider a single isolated normal-conducting region in a superconducting environment. Let us assume for a moment that such an isolated region is thermodynamically stable. In the following, we show that the magnetic flux passing through such a normal conducting region generally can not take arbitrary values.

142

We consider an arbitrary closed loop, which runs in the super-conducting area inside the material. The loop can enclose a normally conductive region as shown in Figure 5.5(a), or the enclosed area is also superconducting. The uniqueness of the wave function, $\psi(\vec{r}) = |\psi(\vec{r})|\,e^{i\Theta(\vec{r})}$, requires the collective phase change along a closed loop $\mathscr{C}$ inside a superconductor to be

$$\oint_{\mathscr{C}} \mathrm{d}\vec{r} \cdot \vec{\nabla}\Theta(\vec{r}) = 2\pi\hat{n}, \tag{5.41}$$

where $\hat{n}$ is an integer number. Since deep inside a superconductor the supercurrent $\vec{j}_s$ is zero, it is possible to derive a relation between the phase change and the magnetic field. From equations (4.78) and (4.79) we find

$$\vec{j}_s(\vec{r}) = q^* n_s^* \vec{v}_s(\vec{r}) = q^* n_s^* \frac{\hbar}{m^*}\left[\vec{\nabla}\Theta(\vec{r}) - \frac{q^*}{\hbar c}\vec{A}(\vec{r})\right] = 0$$

deep inside the superconducting material. Integrating this equation along a closed loop $\mathscr{C}$ inside the superconductor (compare Figure 5.5(a)) and using (5.41), we obtain

$$0 = \oint_{\mathscr{C}} \mathrm{d}\vec{r} \cdot \left(\vec{\nabla}\Theta - \frac{q^*}{\hbar c}\vec{A}\right) = 2\pi\hat{n} - \frac{q^*}{\hbar c}\oint_{\mathscr{C}} \mathrm{d}\vec{r} \cdot \vec{A}. \tag{5.42}$$

The loop integral in the second term can be expressed by a surface integral by use of the Stokes theorem,

$$\oint_{\mathscr{C}} \mathrm{d}\vec{r} \cdot \vec{A} = \int_{\mathcal{S}(\mathscr{C})} \mathrm{d}\vec{f} \cdot \mathrm{rot}\vec{A} = \int_{\mathcal{S}(\mathscr{C})} \mathrm{d}\vec{f} \cdot \vec{B} = \phi.$$

This surface integral over the surface $\mathcal{S}(\mathscr{C})$ enframed by the loop $\mathscr{C}$ describes the magnetic flux ϕ through $\mathcal{S}(\mathscr{C})$. Note that the surface $\mathcal{S}(\mathscr{C})$ may also include a normal conducting region. From (5.42) we find for the magnetic flux the property

$$\phi = \hat{n}\phi_0 \quad \text{with} \quad \phi_0 = \frac{hc}{|q^*|} = \frac{hc}{2e}. \tag{5.43}$$

The important consequence is as follows: The magnetic flux of any surface enframed by a loop inside the superconducting material is *quantized*. It is a multiple of the *flux quantum* ϕ_0. Thus, the magnetic flux is either zero, as it is the case if the material is superconducting inside the loop, or it takes at least the value ϕ_0 if normal conducting regions are present. This result explains the phenomenon of the flux quantization in superconductors mentioned in the section 1.3, which was verified by many experimental works.

5.5.2 – Model of a single vortex

Based on the property of the flux quantization, we first want to construct a model that describes a single normal-conducting region in a superconducting environment without explicitly solving the Ginzburg-Landau equations.

The train of thought on the construction of this model is as follows. We want to describe the phase transition at the lower critical field H_{c1}. There, the pure superconducting state turns into a state in which the magnetic field partially penetrates the material. The idea is that directly at the phase transition individual isolated areas should be formed, in which the normal state is present, and only in these particular regions the magnetic field can penetrate the material. However, we have just learned that the penetrating magnetic field cannot form an arbitrarily large magnetic flux, because the area in which it flows through the material is normally conductive. Rather, an isolated structure will form in which the magnetic flux has a defined value. For such a single structure, we are developing a model in the following.

So we start with the basic idea that a single normal-conducting region forms in such a way that the magnetic flux passing through this region carries one magnetic flux quantum ϕ_0 (or multiples of it). The penetrating magnetic field induces circular supercurrents which flow around this region. This system of circular currents and normal-conducting region is called *vortex*. The normal-conducting region in the center of the vortex is called *vortex core*. We have shown in sections 4.2 and 5.4 that the length scale of significant spatial variation of the superconducting order parameter is given

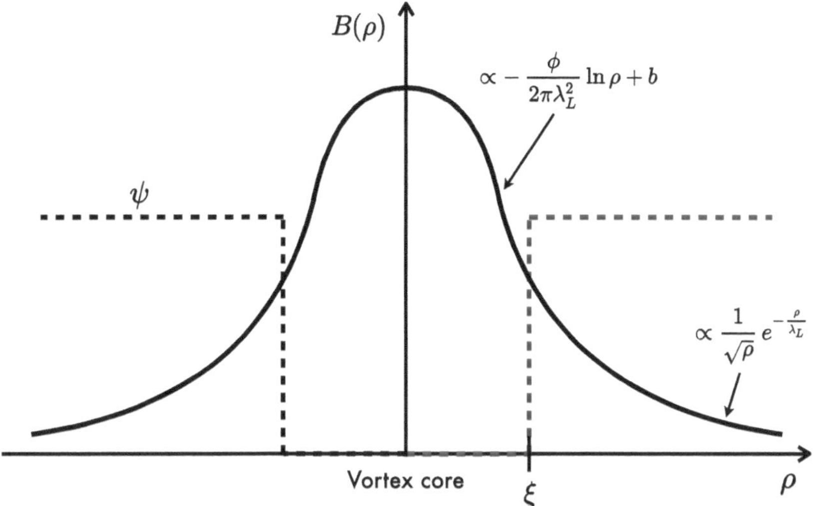

Figure 5.6: Qualitative behavior of the magnetic induction $B(\rho)$ (red solid line) in type 2 superconductors as a function of the distance ρ from the vortex core. Nearby and far away from the vortex the magnetic induction behaves differently. The model for the spatial dependence of the superconducting order parameter ψ is indicated by the blue dashed line.

by the coherence length ξ. Thus, ξ determines also the spatial dimension of the vortex core.

The region in which the magnetic induction changes is determined by the London penetration depth λ_L. We know from section 2.2 that λ_L is also the length scale of the surface currents (leading to the screening of the magnetic field inside the superconducting material). This property can be seen, for example, in Figure 2.1. Thus, also the circular supercurrents of a vortex will decay within this length scale.

Based on these relevant length scales the vortex can be modeled as follows. The basic property of the model is a cylindrical normal-conducting core of radius ξ, centered around the z-axis (Figure 5.5(b)). Within this core, the order parameter is set equal to zero and outside it has the constant value $|\psi_1|$ of the homogeneous solution. The overall course of the order parameter in this model is shown in Figure 5.6 with a dashed line. It can be seen that within the model the core itself is normal conducting but it is surrounded by superconducting material where circular supercurrents (around the z-axis) may flow. These currents produce a magnetic field which forms together with the external magnetic field the field of the magnetic induction.

Now we want to calculate this field of the magnetic induction in the region of the vortex for this model. For this purpose, we use the field equations together with the property of flux quantization.

Let us start with the study of the magnetic induction *outside* the vortex core where the order parameter is finite but constant. The equation that describes the magnetic induction in this region is the third Ginzburg-Landau equation (4.72). When $\psi = \text{const.}$ the first two terms on the right side disappear. In addition, $\vec{\nabla} \times \vec{H} = 0$, since the external magnetic field is homogeneous. So the equation (4.72) becomes

$$0 = \frac{4\pi q^{*2} |\psi_1|^2}{m^* c^2} \vec{A} + \vec{\nabla} \times (\vec{\nabla} \times \vec{A}).$$

With (5.25) one can express the pre-factor by the London penetration depth λ_L. In addition, further formation of the curl on both sides leads us to the equation

$$\vec{\nabla}^2 \vec{B} = \frac{1}{\lambda_L^2} \vec{B},$$

which corresponds to the London equation (2.13). This is the field equation for the region outside the vortex core.

Due to the cylindrical symmetry of the vortex model, it makes sense to move to cylinder coordinates. In these coordinates the magnetic induction takes the form

$$\vec{B} = B_z(\rho)\vec{e}_z,$$

with

$$\rho = \sqrt{x^2 + y^2}\,.$$

Using the Laplace operator in cylindrical coordinates,

$$\vec{\nabla}^2 = \frac{\partial^2}{\partial\rho^2} + \frac{1}{\rho}\frac{\partial}{\partial\rho} + \frac{1}{\rho^2}\frac{\partial^2}{\partial\phi^2} + \frac{\partial^2}{\partial z^2},$$

the London equation reads

$$\frac{d^2 B_z}{d\rho^2} + \frac{1}{\rho}\frac{dB_z}{d\rho} = \frac{B_z}{\lambda_L^2},$$

so that we arrive at a field equation which describes the magnetic induction as a function of the distance ρ to the center of the vortex:

$$\frac{1}{\rho}\frac{d}{d\rho}\left(\rho\frac{dB_z}{d\rho}\right) = \frac{B_z}{\lambda_L^2}. \tag{5.44}$$

This equation corresponds to the London equation for a cylindrical field pointed in z direction.

In the following, we want to use the equation (5.44) to determine the magnetic field for the vortex model. However, to simplify the calculation, we limit ourselves from now on to an *extreme* type 2 superconductor, which is defined by $\kappa \gg 1$. This limitation does not change the physical statement and thus the findings can be easily transferred to all type 2 superconductors.

At first, we derive the relation between the magnetic induction B_z in (5.44) and the total magnetic flux ϕ of one vortex. The total magnetic flux is defined by

$$\phi := 2\pi \int_0^\infty \rho B_z(\rho)\,d\rho\,.$$

It makes sense to decompose the integral into an integration over the vortex core (between 0 and ξ) and an integration over the remaining region outside the vortex core where equation (5.44) is valid. Thus, using (5.44), we may write

$$\phi = 2\pi \int_0^\xi \rho B_z(\rho)\mathrm{d}\rho + 2\pi \lambda_L^2 \int_\xi^\infty \frac{\mathrm{d}}{\mathrm{d}\rho}\left(\rho \frac{\mathrm{d}B_z}{\mathrm{d}\rho}\right)\mathrm{d}\rho\,.$$

We assume that the function $B_z(\rho)$ inside the vortex core changes slowly. In addition, the magnetic induction reaches saturation far away from the core for $\rho \to \infty$ and is therefore also constant there. So $\mathrm{d}B_z/\mathrm{d}\rho = 0$ for $\rho \to \infty$ applies. Under these assumptions the magnetic flux is approximately

$$\phi \approx 2\pi B_z(0)\frac{1}{2}\xi^2 - 2\pi \lambda_L^2 \xi \left.\frac{\mathrm{d}B_z}{\mathrm{d}\rho}\right|_{\rho=\xi}\,.$$

For an extreme type 2 superconductor the first term is very small compared to the second term since $\xi^2/\lambda_L^2 = 1/\kappa^2 \ll 1$. Thus, the relation between the total magnetic flux and magnetic induction of a vortex in an extreme type 2 superconductor is approximately

$$\phi \approx -2\pi \lambda_L^2 \xi \left.\frac{\mathrm{d}B_z}{\mathrm{d}\rho}\right|_{\rho=\xi}\,. \tag{5.45}$$

Field distribution near the vortex core

At first, we study the behavior of the magnetic induction in the proximity of the vortex core. This region may be defined by the range $\xi < \rho \ll \lambda_L$. For $\rho \ll \lambda_L$, the field equation (5.44) reads approximately

$$\frac{\mathrm{d}}{\mathrm{d}\rho}\left(\rho \frac{\mathrm{d}B_z}{\mathrm{d}\rho}\right) \approx 0.$$

This differential equation can be easily integrated. Its solution

$$B_z(\rho) = a \ln \rho + b,$$

where a and b are integration constants, describes the field distribution close to the vortex core. Using equation (5.45), we express the constant a in terms of the constant value ϕ

$$\left.\frac{\mathrm{d}B_z}{\mathrm{d}\rho}\right|_{\rho=\xi} = \frac{a}{\xi} = -\frac{\phi}{2\pi\lambda_L^2\xi} \quad\Rightarrow\quad a = -\frac{\phi}{2\pi\lambda_L^2}.$$

Thus, in the proximity of the vortex core the magnetic field decays logarithmically.

In addition to the magnetic field, we can very easily calculate the distribution of the *supercurrent* around the vortex core. It is obtained by the Ampere's law and the magnetic field distribution known from above,

$$\vec{j}_s = \frac{c}{4\pi}\,\vec{\nabla}\times\vec{B} = \frac{c}{4\pi}\left(-\frac{\mathrm{d}B_z}{\mathrm{d}\rho}\right)\vec{e}_\varphi = -\frac{c}{4\pi}\frac{a}{\rho}\vec{e}_\varphi,$$

where the Nabla operator in cylindrical coordinates was used. From the proportionality $\sim \vec{e}_\varphi$, one can very well see that the supercurrent flows swirling around the vortex core. The current density (and due to the constant particle density also the particle velocity) increases more and more as it approaches the core of the vortex. This looks similar to the wind conditions in a tornado. Furthermore, from the constant a one can recognize that the current density changes like the magnetic field on the length scale λ_L.

In summary, we obtain the following results for the fields $\vec{B}$ and $\vec{j}_s$ in the proximity of the vortex core:

$$\vec{B}(\rho) = \left(-\frac{\phi}{2\pi\lambda_L^2}\ln\rho + b\right)\vec{e}_z, \tag{5.46}$$

$$\vec{j}_s(\rho) = \frac{c\phi}{8\pi^2\lambda_L^2}\frac{1}{\rho}\vec{e}_\varphi. \tag{5.47}$$

Field distribution far away from the vortex core

Next, we study the magnetic induction far away from the vortex core. This region is defined by $\rho \gg \lambda_L$. We use the ansatz

$$B_z(\rho) = B_0 \rho^l e^{-\frac{\rho}{\lambda_L}} \tag{5.48}$$

and calculate the value of l for which this ansatz is a solution of the London equation (5.44). Let us first calculate the following derivatives:

$$\rho \frac{dB_z}{d\rho} = B_0 \left(l\rho^l e^{-\frac{\rho}{\lambda_L}} - \rho^{l+1}\frac{1}{\lambda_L}e^{-\frac{\rho}{\lambda_L}} \right),$$

$$\frac{1}{\rho}\frac{d}{d\rho}\left(\rho \frac{dB_z}{d\rho} \right) = B_0 \left[l^2\rho^{l-2}e^{-\frac{\rho}{\lambda_L}} + l\rho^{l-1}e^{-\frac{\rho}{\lambda_L}}\left(-\frac{1}{\lambda_L} \right) - \right.$$
$$\left. (l+1)\frac{1}{\lambda_L}\rho^{l-1}e^{-\frac{\rho}{\lambda_L}} + \frac{1}{\lambda_L^2}\rho^l e^{-\frac{\rho}{\lambda_L}} \right].$$

According to (5.44), the second equation is equal to B_z/λ_L^2. Using again the ansatz (5.48) and dividing the equation by $B_0\rho^{l-2}e^{-(\rho/\lambda_L)}$, we obtain the following equation,

$$l^2 - (2l+1)\frac{\rho}{\lambda_L} = 0.$$

Far away from the vortex core, where $\rho/\lambda_L \gg 1$, we may neglect the first term l^2 and the above equation can then only be satisfied if $l \approx -1/2$ is fulfilled. Thus, far away from the vortex core, the magnetic induction obeys approximately the following law,

$$B_z(\rho) \approx B_0 \frac{1}{\sqrt{\rho}}e^{-\frac{\rho}{\lambda_L}}.$$

The constant B_0 is fixed by the boundary condition. Consequently, far away from the vortex core, the magnetic field decays exponentially and differs from the logarithmic one described by (5.46) close to the vortex core. The characteristic distribution of $B(\rho)$ is illustrated in Figure 5.6.

The supercurrent follows an exponential law as well. Therefore, far away from the vortex, its value is very small as compared to its value close to the vortex core where (5.47) holds.

Energy of the vortex

Now we want to investigate for which magnetic field strength the vortex model is thermodynamically stable. For this, it is necessary to calculate the energy of a single vortex. More specifically, we are interested in the kinetic energy per unit length caused by the supercurrent flowing around the vortex line. These considerations should lead us to the phase transition at the lower critical point.

In our model it is assumed that the vortex core is metallic, i. e. $|\psi| = 0$ for $\rho < \xi$ (compare the dashed line in Figure 5.6). Furthermore, the order parameter is constant in the proximity of the vortex core, $|\psi| = |\psi_1| = $ const. for $\xi < \rho < \lambda_L$. Far away from the vortex (in the range $\rho > \lambda_L$) we assume that the supercurrent is exponentially negligible. Thus, using the velocity $\vec{v}_s$ of the superconducting 'particles' (superfluid velocity) carrying charge q^* and mass m^*, the kinetic energy per unit length E_V associated with the vortex line reads

$$E_V = 2\pi \int_0^\infty d\rho\, \rho\, \frac{1}{2} m^* |\psi_1|^2 \vec{v}_s^2(\rho) \approx \pi m^* |\psi_1|^2 \int_\xi^{\lambda_L} d\rho\, \rho \vec{v}_s^2(\rho)\,.$$

Finally, we replace $\vec{v}_s(\rho)$ with the supercurrent

$$\vec{j}_s(\rho) = q^* |\psi_1|^2 \vec{v}_s(\rho)$$

according to (4.78) and use the explicit formula (5.47) to evaluate the integral over ρ,

$$E_V \approx \frac{\pi m^*}{|\psi_1|^2 q^{*2}} \int_\xi^{\lambda_L} d\rho\, \rho \vec{j}_s^2(\rho) = \frac{\pi m^*}{|\psi_1|^2 q^{*2}} \frac{c^2 \phi^2}{(8\pi^2)^2 \lambda_L^4} \int_\xi^{\lambda_L} d\rho\, \frac{1}{\rho}$$

$$= \frac{\phi^2}{16\pi^2 \lambda_L^2} \ln \frac{\lambda_L}{\xi}\,, \tag{5.49}$$

where in the last step the known relation

$$|\psi_1|^2 = \frac{m^* c^2}{4\pi q^{*2} \lambda_L^2}$$

was used. Since the magnetic flux has a fixed value according to (5.43), the kinetic energy of a single vortex is also fixed. Therefore, the total energy of the superconducting state in the presence of such vortices is determined by the energy of the superconducting condensate and the number of vortices. We will discuss this in more detail in the next subsection.

5.5.3 – Lower critical magnetic field

The introduced model of a single vortex is now to be used to describe the inhomogeneous superconducting state near the lower critical field H_{c1}. In particular, the value of H_{c1} should be calculated as a function of the system parameters.

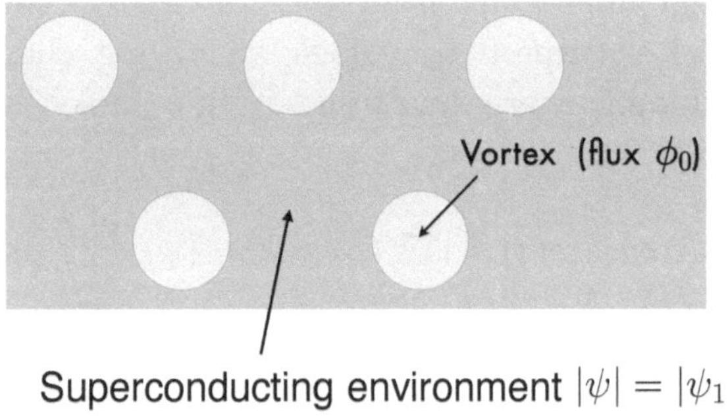

Figure 5.7: Model of independent vortices (light gray) near the lower critical magnetic field H_{c1}. In the superconducting region (dark gray) the order parameter is assumed to be equal to the value $|\psi_1|$ of the homogeneous solution. Each vortex carries one flux quantum ϕ_0.

Again, let us start from a few assumptions and simplifications. The assumptions are as follows. (i) At a certain critical magnetic field the superconductor forms a system of vortices, where each vortex is described by the vortex model introduced above. Let this system of vortices be characterized by $n = N/A$ vortex lines per unit area, each one carrying one flux quantum $\phi_0 = 2\pi\hbar c/|q^*|$ which is the lowest possible non-zero flux value per vortex (compare subsection 5.5.1). (ii) Outside the vortices, the superconducting order parameter is modeled by its constant value of the homogeneous solution, $|\psi| = |\psi_1|$. (iii) For field values close to the lower critical field H_{c1} we may assume that the vortex density is sufficiently small, so that the average distance between the vortices is much larger than the London penetration depth, $n^{-1/2} \gg \lambda_L$, and the vortices can be treated independently. The model of independent vortices just described is illustrated in Figure 5.7.

We consider again an extreme type 2 superconductor characterized by $\kappa \gg 1$ in an homogeneous external magnetic field

$$\vec{H} = H\vec{e}_z.$$

The lower critical point H_{c1} is determined by the equality $G_1 = G_3$ of the Gibbs free energies of homogeneous and inhomogeneous state (compare the intersection point at H_{c1} in Figure 5.4(b)). So in order to calculate the value H_{c1}, we have to consider the Gibbs free energy for the introduced vortex model. According to (3.22), the Gibbs free energy is defined by the equation,

$$G[\vec{H},\psi] = F[\vec{B},\psi] - \frac{1}{4\pi}\int_{\mathcal{V}} d^3r\, \vec{B}(\vec{r}) \cdot \vec{H}. \tag{5.50}$$

All material properties are symmetric in z direction so that our problem reduces to a two-dimensional treatment. Furthermore, according to our model, we assume that the magnetic induction penetrates into the superconducting material by forming N independent vortex lines, each one carrying one flux quantum ϕ_0 (compare Figure 5.7). Thus, within our model, the volume integral of the second term in the Gibbs free energy (5.50) reduces to a surface integral,

$$\int_{\mathcal{V}} d^3r\, \vec{B}(\vec{r}) \cdot \vec{H} = LH\int_{\mathcal{S}} df\, B_z(\vec{r}) = LHN\phi_0,$$

where a surface $\mathcal{S}$ perpendicular to the vortex lines is chosen. The length L denotes the height (in z direction) of the superconducting material.

Introducing the area $A = V/L$ perpendicular to the z direction, we find from the defining equation (5.50) for the Gibbs free energy of the vortex system (per volume V),

$$\frac{G[\vec{H},\psi]}{V} = \frac{F[\vec{B},\psi]}{V} - \frac{H}{4\pi} n\phi_0, \qquad (5.51)$$

where $n = N/A$ is the number of vortex lines per unit area. Note that $B(n) = n\phi_0$ can be considered as the average magnetic induction in the sample. Further note that G in the expression (5.51) corresponds of the Gibbs free energy G_3 in the inhomogeneous state.

For further evaluation of the expression (5.51), one needs the free energy F of the system. According to (4.49), the free energy of the total volume can be decomposed into four parts,

$$F = F_{M,n} + F_{vor} + F_{bulk} + F_{EM}, \qquad (5.52)$$

where

$$F_{vor} = \int_{\mathcal{V}} \mathrm{d}^3 r \, \frac{1}{2m^*} \left| \left(\frac{\hbar}{i}\vec{\nabla} - \frac{q^*}{c}\vec{A}(\vec{r}) \right) \psi(\vec{r}) \right|^2,$$

$$F_{bulk} = \int_{\mathcal{V}} \mathrm{d}^3 r \left[r |\psi(\vec{r})|^2 + \frac{u}{2} |\psi(\vec{r})|^4 \right],$$

$$F_{EM} = \frac{1}{8\pi} \int_{\mathcal{V}} \mathrm{d}^3 r \, \vec{B}^2(\vec{r}).$$

In the following, we want to interpret all the above parts of the free energy one after the other. We start with the first part F_{vor}. One can see that the corresponding expression contains the momentum operator

$$\mathscr{P} = \frac{\hbar}{i}\vec{\nabla} - \frac{q^*}{c}\vec{A}$$

known from the replacement (4.45). If the field $\psi(\vec{r})$ is treated as the wave function of the superconducting condensate, then the energy part F_{vor} can be written in a compact form in terms of the expectation value $\langle \mathscr{P}^2 \rangle$ as follows,

$$F_{vor} = \frac{1}{2m^*} \langle \mathscr{P}^2 \rangle .$$

Thus, F_{vor} has to be interpreted as the total kinetic energy of the superconducting 'particles'. Within our model the kinetic energy is determined by the energy E_V of the supercurrent flowing around a single vortex. We have already calculated this energy E_V and the result was given by equation (5.49). Thus, F_{vor} describes the kinetic energy E_V per vortex and unit length, as given by equation (5.49), times the number N and height L of the vortices, $F_{vor} = NLE_V$. Using again the number $n = N/A$ of vortex lines per unit area, we obtain

$$F_{vor} = nVE_V . \tag{5.53}$$

The quantity F_{bulk} is the part of F caused by the presence of the superconducting order parameter ψ. Near the transition point, the density of vortices is very small. In this case, $|\psi(\vec{r})|$ can be replaced approximately with the constant value $|\psi_1|$ outside the vortex cores and we can write

$$F_{bulk} \approx V \left(r|\psi_1|^2 + \frac{u}{2}|\psi_1|^4 \right) . \tag{5.54}$$

Using the same approximation also the field part F_{EM} can be expressed by the average magnetic induction $B(n) = n\phi_0$,

$$F_{EM} \approx \frac{B^2(n)}{8\pi}V = \frac{n^2\phi_0^2}{8\pi}V . \tag{5.55}$$

Finally, replacing in (5.52) the different contributions with (5.53), (5.54), and (5.55), we obtain for the Gibbs free energy (5.51) per volume the approximate expression,

$$\frac{G(H,\psi,n)}{V} \approx g_0 + nE_V + r|\psi_1|^2 + \frac{u}{2}|\psi_1|^4$$

$$+\frac{1}{8\pi}n^2\phi_0^2 - \frac{H}{4\pi}n\phi_0, \tag{5.56}$$

where $g_0 = F_{M,n}/V$.

The above expression (5.56) describes the Gibbs free energy of the superconducting system as a function of the vortex density n. Due to the negative sign of the last term, it may happen that — depending on the value of the external field H — the Gibbs free energy becomes smaller when the vortex density increases. Such a situation leads to an instability towards the formation of vortices.

Using this expression for the Gibbs free energy, we now want to calculate the lower critical value H_{c1}. This should correspond to the field strength at which the first vortices form and the magnetic field begins to penetrate the material. According to the Figure 5.4(b), the transition at H_{c1} is defined by the fact that there is no difference between G_3 (state with vortices) and G_1 (state without vortices). Thus, within our vortex model one defines the lower critical field H_{c1} as the particular field value at which the formation of vortices (or at least one vortex) leads to the same Gibbs free energy as for the pure superconductor without any vortices ($n = 0$). Consequently, if the Gibbs free energy as a function of the vortex density n is known, a reasonable definition of the lower critical field can be formulated as follows,

$$\left.\frac{\partial G}{\partial n}\right|_{n=0,H=H_{c1}} = 0. \tag{5.57}$$

The partial derivative with respect to n is easily evaluated from expression (5.56). For arbitrary field values H, it reads

$$\left.\frac{\partial G}{\partial n}\right|_{n=0} \approx V\left(E_V + \frac{1}{4\pi}n\phi_0^2 - \frac{H}{4\pi}\phi_0\right)_{n=0} = V\left(E_V - \frac{H}{4\pi}\phi_0\right).$$

Thus, for $H > (4\pi/\phi_0)E_V$, the formation of vortices becomes energetically favorable,

$$\left.\frac{\partial G}{\partial n}\right|_{n=0} < 0 \quad \text{for} \quad H > 4\pi\frac{E_V}{\phi_0}\,.$$

In this case, G_3 is smaller than G_1 and in Figure 5.4(b) the magnetic field is to the right of the first intersection point at H_{c1}. From equation (5.57) we get the determination equation for H_{c1},

$$V\left(E_V - \frac{H_{c1}}{4\pi}\phi_0\right) = 0.$$

From this equation it follows with (5.49) and $\phi = \phi_0$:

$$H_{c1} = 4\pi\frac{E_V}{\phi_0} = 4\pi\frac{\phi_0}{16\pi^2\lambda_L^2}\ln\frac{\lambda_L}{\xi} = \frac{\phi_0}{4\pi\lambda_L^2}\ln\frac{\lambda_L}{\xi}\,. \tag{5.58}$$

Since the flux quantum according to (5.43) is determined by natural constants alone, the result (5.58) can be used to calculate the value H_{c1} from the knowledge of the system parameters ξ and λ_L. This is of fundamental importance for experiments, as these parameters are easily accessible.

From a theoretical point of view, however, it is still important to compare the above result for H_{c1} with the critical magnetic field H_c of the ordinary transition into the homogeneous superconducting state. According to our qualitative considerations from section 5.4, H_{c1} should be smaller than H_c. To do this, we replace the flux quantum ϕ_0 with the parameter H_c. We start from equation (5.13) which can be solved for H_c:

$$H_c = \frac{1}{\sqrt{2}}\frac{\hbar c}{q^*\xi\lambda_L}\,.$$

Replacing the charge q^* with the flux quantum $\phi_0 = hc/q^*$ and $\hbar = h/2\pi$, we find for H_c,

$$H_c = \frac{1}{\sqrt{2}}\frac{\phi_0}{2\pi\xi\lambda_L} = \frac{\phi_0}{4\pi\lambda_L^2}\sqrt{2}\frac{\lambda_L}{\xi}\,. \tag{5.59}$$

Finally, we insert this expression into (5.58) and obtain the final result for the lower critical field H_{c1} of an extreme type 2 superconductor, formulated in terms of H_c and the Ginzburg-Landau parameter $\kappa = \lambda_L/\xi$,

$$H_{c1} = \frac{H_c}{\kappa\sqrt{2}} \ln \kappa \quad \text{for} \quad \kappa \gg 1.$$

In agreement with our qualitative considerations of section 5.4 we find that the critical field value H_{c1} is indeed *smaller* than the material parameter H_c. Our model explains this transition in such a way that when the external magnetic field reaches H_{c1}, the material goes into a new state which is characterized by isolated vortices in the superconducting environment. Thereby, close to the transition point, each of these vortices is described by the introduced vortex model. Each vortex carries one flux quantum ϕ_0, and the several vortices have to be treated as independent systems. It also follows that if the field strength H is increased, the number of vortices also increases.

Finally, we should emphasize that our model is not able to draw any conclusions about the spatial position of the vortices. In real materials, vortices are strongly influenced by impurities or defects of the material. In this context, one speaks about 'pinning' of vortices. Note that the description within our vortex model is restricted to the regime close to the lower transition point H_{c1} where the vortex density is small enough to treat the vortices as independent systems.

5.5.4 – Upper critical magnetic field

Next, we turn to the discussion of the behavior close to the second transition point H_{c2} where the material undergoes the transition from the inhomogeneous state to the normal metal. In Figure 5.4(b), this is the intersection point of the G_3 curve with the G_2 curve, which is placed right of H_c. According to our vortex model, near this transition point, the density of the vortices has increased so much that they begin to overlap and can no longer be treated as independent systems. Therefore, the vortex model is not applicable anymore for the quantitative description of the transition. Instead, we use a different approach which is based on the following idea.

Due to the overlap of vortices it can be assumed that near this second transition point, the superconducting order parameter has greatly reduced on average. Consequently, we can assume that in this region the limiting case $|\psi_3(\vec{r})| \ll 1$ applies to the inhomogeneous solution. Thus, our starting point for the following considerations is to linearize the Ginzburg-Landau equations and to study the condition for possible superconducting solutions on the basis of a linear system of equations.

The first step is again the simplification of the discussion by considering a homogeneous magnetic field $\vec{H} = H\vec{e}_z$. If the order parameter ψ is to be small near the transition point H_{c2} then all screening currents $j \propto |\psi|^2$ are small as well and the magnetic induction $\vec{B}$ is approximately equal to $\vec{H}$ (in lowest order),

$$\vec{B} = B\vec{e}_z \approx \vec{H}.$$

We choose the vector potential

$$\vec{A} = Bx\vec{e}_y, \tag{5.60}$$

which leads to $\vec{B} = \mathrm{rot}\,\vec{A} = B\vec{e}_z$. This is already the solution for the magnetic induction after the linearization.

The next step is the linearization of the Ginzburg-Landau equation for the order parameter. It proves to be favorable to start from the form (4.59),

$$0 = -\frac{\hbar^2}{2m^*}\left(\vec{\nabla} - \frac{iq^*}{\hbar c}\vec{A}\right)^2 \psi + r\psi + u|\psi|^2\psi.$$

Within the linearization we may neglect the third order term $\propto \psi^3$ and the approximate Ginzburg-Landau equation which is valid close to the upper transition point reads

$$0 \approx -\frac{\hbar^2}{2m^*}\left(\vec{\nabla} - \frac{iq^*}{\hbar c}\vec{A}\right)^2 \psi + r\psi.$$

In the following, we want to solve this linear differential equation in the presence of the vector potential (5.60). After an expansion of the differential operator, it reads

$$0 = -\frac{\hbar^2}{2m^*}\left[\vec{\nabla}^2\psi - \frac{iq^*}{\hbar c}\vec{\nabla}\cdot(Bx\psi\vec{e}_y) - \frac{iq^*}{\hbar c}Bx\vec{e}_y\cdot\vec{\nabla}\psi\right.$$
$$\left. -\frac{q^{*2}}{\hbar^2 c^2}B^2x^2\psi\right] + r\psi = -\frac{\hbar^2}{2m^*}\left[\vec{\nabla}^2\psi - 2\frac{iq^*}{\hbar c}Bx\frac{\partial\psi}{\partial y}\right.$$
$$\left. -\frac{q^{*2}}{\hbar^2 c^2}B^2x^2\psi\right] + r\psi.$$

The result can be written in the form

$$\left[-\frac{\hbar^2}{2m^*}\vec{\nabla}^2 + \frac{m^*}{2}\omega_c^2 x^2 - i\hbar\omega_c\, x\frac{\partial}{\partial y}\right]\psi = |r|\psi, \qquad (5.61)$$

with

$$\omega_c = \frac{|q^*|B}{m^*c}. \qquad (5.62)$$

Equation (5.61) is the Schrödinger equation of charged particles (charge $q^* = -2e$) in an external magnetic field. As is known from quantum theory, the particles can only occupy orbits with discrete energy values, called Landau levels. The so-called cyclotron frequency ω_c describes the energy difference between two neighboring Landau levels.

Since equation (5.61) is a linear differential equation with respect to the variable ψ, we can make an exponential ansatz for the solution function. An appropriate ansatz reads

$$\psi(\vec{r}) = e^{ik_y y + ik_z z}f(x),$$

which leads us to a differential equation for the function $f(x)$,

$$-\frac{\hbar^2}{2m^*}\left(f'' - k_y^2 f - k_z^2 f\right) + \frac{m^*}{2}\omega_c^2 x^2 f + \hbar\omega_c\, xk_y f = |r|f.$$

We bring this equation in an appropriate form,

$$-\frac{\hbar^2}{2m^*}f'' + \left(\frac{m^*}{2}\omega_c^2 x^2 + \hbar\omega_c k_y x + \frac{\hbar^2}{2m^*}k_y^2\right)f$$

$$+ \frac{\hbar^2}{2m^*}k_z^2 f = |r|f,$$

which can be further simplified to

$$-\frac{\hbar^2}{2m^*}f'' + \frac{m^*}{2}\omega_c^2\left(x + \frac{\hbar k_y}{m^*\omega_c}\right)^2 f + \frac{\hbar^2}{2m^*}k_z^2 f = |r|f$$

or

$$-\frac{\hbar^2}{2m^*}f'' + \frac{m^*}{2}\omega_c^2(x + x_0)^2 f = \left(|r| - \frac{\hbar^2 k_z^2}{2m^*}\right)f.$$

This is the Schrödinger equation of a one-dimensional harmonic oscillator with a minimum of the potential at a certain place x_0 shifted from the origin. The value of x_0 is

$$x_0 = \frac{\hbar k_y}{m^*\omega_c} = \frac{\hbar k_y c}{|q^*|B}.$$

From a comparison with the solution of the Schrödinger equation for an harmonic oscillator, as known from quantum theory, we immediately obtain the quantized energy values (eigen-energy values),

$$\hbar\omega_c\left(\hat{n} + \frac{1}{2}\right) = |r| - \frac{\hbar^2 k_z^2}{2m^*}, \tag{5.63}$$

where $\hat{n} = 0, 1, 2, \ldots$ are integer numbers.

We now want to move on to making statements about the phase transition to the normal state using the equation (5.63). In particular, we can show that there is a maximum value of the magnetic field — per definition the upper critical field H_{c2} — where superconducting solutions are possible. This upper critical value can be calculated from the eigen-value equation (5.63) as follows.

With the explicit expression (5.62) for ω_c, the above equation (5.63) can be written in the form

$$|r| = \hbar\frac{|q^*|B}{m^*c}\left(\hat{n} + \frac{1}{2}\right) + \frac{\hbar^2 k_z^2}{2m^*}. \tag{5.64}$$

The parameter $|r|$ on the left is fixed and independent of B. It only depends on the temperature. We assume the equality

$$B = H,$$

which is valid near the transition point (ψ small). If an external magnetic field H is applied, the magnetic induction B is predetermined (due to $B = H$) and enters on the right side of the equation (5.64). Now the still free variables $\hat{n}$ and k_z are set in such a way that with given B and $|r|$ the equation (5.64) is fulfilled.

One can easily see from the structure of equation (5.64) that the equation cannot be fulfilled for arbitrarily large values of B. The largest possible value of B is determined by $\hat{n} = 0$ and $k_z = 0$. This is due to the fact that all terms of the equation are positive. If this maximum value of B is applied, the equation can no longer be fulfilled by non-zero values for $\hat{n}$ and k_z, as this would make the right side too large. This leads to the following determination equation of the upper critical field,

$$|r| = \frac{\hbar|q^*|H_{c2}}{2m^*c}.$$

For $H > H_{c2}$, equation (5.64) can not be fulfilled for any (n, k_z). In this case no superconducting solution exists.

In terms of $\xi^2 = \hbar^2/(2m^*|r|)$ and $\phi_0 = hc/|q^*|$, the upper critical field reads

$$H_{c2} = \frac{2m^*c|r|}{\hbar|q^*|} = \frac{\hbar c}{|q^*|\xi^2} = \frac{\phi_0}{2\pi\xi^2}. \tag{5.65}$$

Thus, at the upper critical field, a cylinder of radius ξ contains one half of a flux quantum. Again, one can see that H_{c2} is determined by system parameters (in this case only the coherence length), which are experimentally easy to determine.

Finally, let us again compare the obtained value of H_{c2} with the parameter

$$H_c = \sqrt{4\pi r^2/u}.$$

Using (5.59), we replace in (5.65) ϕ_0 with H_c,

$$H_{c2} = \frac{1}{2\pi\xi^2}\phi_0 = \frac{1}{2\pi\xi^2}\sqrt{2}2\pi\xi\lambda_L H_c = \sqrt{2}\frac{\lambda_L}{\xi}H_c.$$

Expressed in terms of the Ginzburg-Landau parameter $\kappa = \lambda_L/\xi$, the upper critical field takes the compact form

$$\boxed{H_{c2} = \kappa\sqrt{2}H_c.}$$

For a type 2 superconductor characterized by $\kappa > 1/\sqrt{2}$ the upper critical field H_{c2} is larger than the thermodynamic critical field H_c of the corresponding homogeneous transition.

Since the superconducting order parameter continuously disappears during this transition, H_{c2} is a second-order phase transition. The fact that when the equation (5.64) is solved, there is no critical field strength *below* H_c means that the phase transition at H_{c1} cannot be of second order. Instead, the transition at H_{c1} must be of first-order.

5.5.5 – Magnetic induction versus magnetic field

We now summarize the results from the previous subsections to discuss the general behavior of type 2 superconductors with respect to a variation of the external magnetic field H. The magnetic induction $B(H)$ as a function of H, as measured at a representative place inside the superconducting material (for example near a vortex), is qualitatively shown in Figure 5.8.

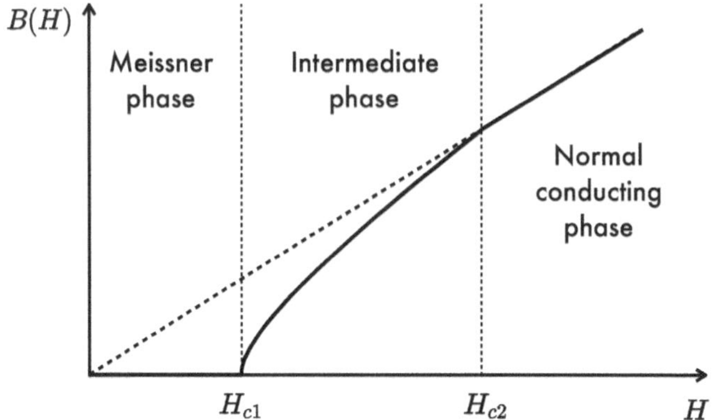

Figure 5.8: Qualitative behavior of the magnetic induction B inside a type 2 superconductor as a function of the external magnetic field H. Between the Meissner phase characterized by $B \approx 0$ and the normal conducting phase where $B \approx H$ an intermediate phase is found. The intermediate phase is characterized by partial penetration of the magnetic field in form of a system of vortices.

At small field values below the lower critical field, $H < H_{c1}$, the system is in the Meissner phase. Here the magnetic induction is suppressed, as discussed in section 2.2. If the external magnetic field reaches its lower critical value

$$H_{c1} \approx (H_c/\kappa\sqrt{2})\ln\kappa$$

(for $\kappa \gg 1$), then the penetration of the magnetic induction is energetically favored by forming a system of independent vortices. The corresponding phase transition is of first order. This phase is present in a certain range of field values

$$H_{c1} < H < H_{c2}.$$

Each single vortex carries one flux quantum ϕ_0 and the density of such vortices increases with increasing magnetic field H.

If H is close to the upper critical field value

$$H_{c2} = \kappa\sqrt{2}H_c,$$

the vortex cores begin to overlap and the superconducting region disappears at H_{c2}. The phase transition here is of second order, i. e. the superconducting order parameter disappears continuously. This allows a linearization of the Ginzburg-Landau equations. The general solution of the differential equation (5.61), which is valid for field values very close to the upper critical field, can be found by an appropriate ansatz. It leads to the well-known Abrikosov flux lattice (Abrikosov 1957, Nobel Prize 2003). In contrast to the situation near the lower critical field, where the vortices are rather disordered, the state near H_{c2} is characterized by a hexagonal (ordered) lattice of vortices. The Abrikosov flux lattice is schematically shown in Figure 1.6.

The solution of the linearized Ginzburg-Landau equation for H very close to H_{c2} can be found exactly (not shown here) and leads to the Abrikosov lattice. In general, however, the non-linear Ginzburg-Landau equations have to be solved numerically.